The ABCs of Safe Flying

The ABCs of Safe Flying

4th Edition

David A. Frazier

McGraw-Hill

New York   San Francisco   Washington, D.C.   Auckland   Bogotá
Caracas   Lisbon   London   Madrid   Mexico City   Milan
Montreal   New Delhi   San Juan   Singapore
Sydney   Tokyo   Toronto

McGraw-Hill

A Division of The **McGraw·Hill** *Companies*

2 3 4 5 6 7 8 9 0   DOC/DOC   9 0 3 2 1 0 9

ISBN 0-07-021997-4 (PBK)
     0-07-021996-6 (HC)

The sponsoring editor for this book was Shelley Ingram Carr, the editing supervisor was Sally Glover, and the production supervisor was Pamela Pelton. It was set in Times by Patricia Caruso of McGraw-Hill's Professional Book Group Composition Unit, Hightstown, NJ.

Printed and bound by R. R. Donnelley & Sons Company.

McGraw-Hill books are available at special quantity discounts to use as premiums and sales promotions, or for use in corporate training programs. For more information, please write to the Director of Special Sales, McGraw-Hill, 11 West 19th Street, New York, NY 10011. Or contact your local bookstore.

 This book is printed on recycled, acid-free paper containing a minimum of 50 percent recycled, de-inked fiber.

PFS

To Welby and Ella Harvey:
Your support, encouragement, and love throughout
the years helped make me whatever I am today.
And, oh yes, I believe your third daughter
had a teeny bit of positive influence, too.

Contents

CONTENTS

10   Federal Aviation Regulations (Changes To)   *189*

11   Aviation Careers   *197*

Glossary   *215*

Index   *225*

About the Author   *232*

Acknowledgments

My thanks to the Cessna Aircraft Company, Boeing Aircraft Company, McDonnell-Douglas Aircraft Company, Tim Litzer, Vince, Vicki, and Janene for their help and support. My special thanks to my beloved Janene, without whose help in all areas of my life, I would be lost.

Introduction

PILOT IS A NOUN DEFINED BY WEBSTER AS "ONE WHO STEERS A SHIP OR AIRCRAFT." Of the thousands of words and definitions given in the dictionary, the word "*pilot*" is one of the most oversimplified. Could not a child steer a ship or aircraft? Of course, but I think the end result would be predictable. Are you a pilot? Or, do you fly?

Fly is a verb Webster defines as "to move through the air with wings, as a bird." Now that's more like it. Right? It has such a magical sound. "To move through the air with wings, as a bird." If only we could, but we can't without the aid of an aircraft.

INTRODUCTION

Aircraft is defined as "any machine designed for air navigation under its own power." An airplane, glider, helicopter, or dirigible balloon would qualify under the category of aircraft. And even though the material discussed in this book most likely is adaptable to all, we will be thinking in terms of the airplane.

Now, grouping these three words into a simple sentence, you begin to form a more complete understanding of the overall objective. "Pilots fly aircraft." Instead of a list of singular perceptions each with its own definition, the three words are grouped together to form a meaningful sentence that gives us insight. Insight is the highest form of learning. It is the association and correlation of a group of singular perceptions into a meaningful block of knowledge.

Learning to fly proceeds in exactly the same manner. You must start at the beginning and learn. You learn one step at a time, from the simple to the complex, from the known to the unknown. This method of learning is called the *building-block technique* of instruction.

Actually, the building-block technique is a very apt name for how people learn. Think for a moment of the great pyramids of Egypt, ancient monuments to human ingenuity, handiwork, and symmetry. They are an engineering marvel that might never again be equaled by human endeavor. But what has kept the pyramids standing for all these thousands of years? It's simple. The pyramids have a firm foundation upon which to place the giant stone blocks, one at a time, each in its own place, and each in itself a key to the strength of the structure. If one block had been left out or placed out of sequence, these giants would have been reduced to rubble many years ago.

Learning to fly is a task that should be undertaken in exactly the same manner as the building of the pyramids. Take it one step at a time, in the proper order, and build knowledge upon knowledge. To be a really good pilot, there are no shortcuts. If you leave something out of your training or your instructor does not demand that you learn in a proper sequence, your training and resultant piloting abilities will be less than they should be. Then, as a pyramid with a stone left out, your proficiency will erode to the point where you will be a danger to yourself and any other person flying with you, near you, or under you.

The FAA and National Transportation Safety Board files are full of statistics of needless airplane accidents and incidents attributed to pilot error. Many of these could have been avoided with proper training and practice. But the accident I'm most interested in preventing is *yours*.

Over the years I have found that the best way to prevent a needless accident, or incident, is to utilize a method I call the ABC method. This stands for *attitude, basics,* and *communication.* The ABC method can be taken in many ways. For instance, attitude can be the attitude of the aircraft, the mental attitude of the pilot, the attitude of the instructor, or an overall attitude between the student and instructor. The basics can be a basic flight technique, basic planning, or just good old basic common sense. Communication can be the innermost thoughts of the pilot, communication with the instructor, tower, FSS, ground crew, or any other manner of getting a point across.

To illustrate my point, come back with me to a small fixed-base operation in Texas, where a number of years ago I started as a flight instructor.

The incident started out as most incidents do—quietly, innocently, and almost imperceptibly. Its end could have been tragic.

At this operation, which doubled as a flight school and crop-dusting service, we had about 15 aircraft, 50 or 60 students, 5 flight instructors, and 2 mechanics. In this particular instance, the mechanics provided the tragic comedy.

One mechanic was mature, experienced, and confident, and the other was young, green, and equally as confident. Out behind the repair shop sat an old Piper Pawnee that hadn't flown for years. The mechanics had a running verbal battle as to whether or not the old Pawnee would run. The older mechanic said he could get the engine in shape and have it running in a few days. The younger mechanic thought the best idea would be to remove the prop, lay it aside, and insert a new airplane behind it. It was entertaining to go out to the shop and start an immediate argument.

There finally came a slack period and into the shop came the old Pawnee. Bets were being laid down as if it were Super Bowl Sunday. The two mechanics began removing the chemical tank and tearing down the engine. They poured some type of lubricant into the cylinders and turned the engine through for hours.

Throughout the process, the older mechanic was confidently quiet. His younger counterpart, on the other hand, was constantly laughing. He considered the effort to be pointless busywork because he believed the engine would never run. The two did work very hard on the engine, but it was the exclusion of some other important details that they would later regret.

With the day of truth at hand, we all watched in anticipation as they backed the old duster from the hangar. They turned it 90 degrees from the open shop door and pointed it directly at our brand new twin Commander, which happened to be parked about 50 feet away. After a last-minute exchange of insults, the older mechanic climbed into the cockpit. (The younger mechanic refused to get in, saying it would be a waste of his energy because nothing was going to happen.) The young mechanic walked around and leaned casually on the wing nearest the open shop door. He nearly died laughing as the older mechanic opened the fold-down doors of the Pawnee and told us to clear the prop.

Assured it was clear, he hit the starter. No more than two blades had gone by when the engine roared to life. (It seems that in their haste, a cable in the throttle connection had been hooked up backwards, and although it appeared to be at idle, the throttle was in fact nearly full open.)

Huge grey-black clouds of smoke poured back from the engine through the now vacant chemical tank housing and instantly filled the cockpit. It overflowed through the open doors of the now moving aircraft and was blown almost straight back by the prop blast.

The young mechanic instinctively grabbed the left wing and hung on. This caused the Pawnee to make an almost perfect 90-degree left turn. As the aircraft turned, it ran over a water pipe that was sticking up from the ground. The pipe broke off at the ground and water shot straight up 50 feet in the air.

During the split-second it took for this to transpire, the only detectable action from the smoke-filled cockpit was the older mechanic's arms thrashing wildly, apparently attempting to grasp some unseen anchor.

INTRODUCTION

As we all stood, frozen and awestruck, the aircraft reached the completion of its turn and crashed through the metal side of the shop building. When it reached the wingroot, it was over.

As the smoke began to clear, with water still shooting high into the air, the older mechanic slowly climbed from the cockpit of the very broken Pawnee, which was firmly implanted in the side of the shop. He didn't say a word. He walked rather swiftly around the back of the plane and went directly up to the young mechanic. He put his nose about three inches from the face of the young man and screamed something I'll never forget if I live a thousand years. He said, "I told you the s.o.b. would start!"

Although this incident was undoubtedly the funniest thing I have ever witnessed, it was at the same time potentially disastrous. There were many people standing around watching, and a full-throttle runaway airplane pulled by a six-foot steel prop is not something I care to challenge.

Everyone was lucky, since the only damage was to the plane and the shop. The already old Pawnee was now a basket case, and the shop was left with a gaping hole that had not been in the blueprints.

It was funny and nobody was hurt, but it should not have happened. The people involved did not even come close to using the ABCs. They were prime for an accident from the very beginning. Their attitude was one of "I'll show you." Well, I guess they did show each other something. If nothing else, they showed a complete disregard for planning and preparation. They were both so busy trying to look good in the eyes of everyone else that their attitudes slipped into a pattern of "I'm right and I can do it." Dangerous, as well as stupid.

They also threw away many basic precepts of safety and good sense. The plane was not tied down or chocked. They didn't even move the plane away from people and buildings. The brakes were not checked. (They didn't have enough fluid in them to stop a runaway skateboard.) The man in the cockpit was not qualified to be there. He couldn't find the throttle, mags, mixture, or fuel selector to shut down the runaway engine. In addition, he apparently had not been well schooled in emergency procedures. (They don't all happen at 6,000 feet.) He panicked—a common, yet often dangerous human reaction. If a person is sound in basic flight technique, this reaction can be forced way down on the list of things to do. In other words, if everything else, including prayer, has failed, then maybe it's time to panic. But it should not be the first thing to occur.

And what a lack of communication! The only true communication that occurred during this incident was a constant verbal assault on each other's intelligence and the clearing of the prop (at least something was done correctly).

These men should have discussed their intentions and formed a plan of action. They should have made a safety checklist and followed it. Not having a checklist for something as unpredictable as this venture was foolish. They didn't even have it tied down!

This brings me to another related point explored in more depth later in this book—they were in a hurry. Each one was so intent on proving the other wrong that haste overtook good sense. It has killed far too many otherwise fine pilots. I hurry. You hurry. We all hurry. But hurrying has few places in safe aviation. Slow down, think it over, and plan it out. Then, if anything goes wrong, you probably will have a *correct* plan of action.

Hence, always keep in mind the ABCs—*attitude, basics,* and *communication.*

1
The Total Preflight

ACCORDING TO REPORTS FROM THE FEDERAL AVIATION ADMINIS-
tration, the National Transportation Safety Board, and the National Aeronautics and Space Administration, the single most hazardous element in aviation is the pilot. That's right—the pilot.

THE PILOT

They have very sterile wording to sum up the probable causes of aircraft accidents/ incidents: inadequate preflight planning/preparation; attempted operation beyond experience/ability level; or improper in-flight decisions or planning. These are all very polite ways of saying the same thing—the pilot blew it.

The sad part of all this is that most aircraft accidents could have been and can be avoided. Only a very small portion is caused by mechanical failure. The rest are pilot-induced through overconfidence, ignorance, or a lackadaisical approach to their flight technique. In short, they ignore the ABCs.

Consider this random selection from the files of the NTSB. It is File # 3-3851. On November 12, 1996, near Wise, North Carolina, a Piper PA28 departed from a cloud-bank with a portion of one wing missing. The aircraft crashed in an uncontrolled descent, and the pilot and three passengers were killed.

Chapter One

As is the case with many aircraft accidents, the exact cause will never be known. What *is* known is that the aircraft departed Sussex, New Jersey, enroute to Myrtle Beach, South Carolina, with the pilot and three passengers on board. The pilot, age 37, had 100 hours total time with only five hours in type. He was not instrument rated. No known weather briefing was obtained prior to takeoff. An in-flight forecast was received that called for marginal VFR (if not IFR) weather conditions. The flight continued until the aircraft impacted the ground at 1313 local time.

The weather at the crash site was 600 feet overcast, four miles visibility with fog and drizzle (definitely IFR). The NTSB lists the probable causes as operation beyond experience/ability level; spatial disorientation (vertigo); exceeded stress limits of the aircraft, resulting in aircraft separation in flight. Type of flight plan—none.

It was a blueprint for a tragedy. A relatively inexperienced, noninstrument-rated private pilot departed into known IFR weather with no briefing or flight plan. Result—four needless deaths.

Where would the prevention of this type of accident start? With the pilot. A noninstrument-rated private pilot has about as much business attempting a four-state cross-country flight in marginal weather as I would racing at Indianapolis. The odds are about the same (believe me, I have absolutely no business racing at Indianapolis). The end results are quite predictable in both instances—tragedy. If you are a person who likes to play the odds and gamble with life, go ahead and gamble with your own, but don't take innocent people with you.

Maybe the people in the aforementioned case knew what they were getting into, but I suspect they didn't. The average passenger places blind trust in the pilot of an aircraft. Often, this trust is not warranted. A pilot's license does not automatically make someone the keeper of all wisdom. It means that he, at the time of the checkride, possessed the *minimum* qualifications for the license or rating. Maybe the skills have not been further developed, or even worse, maybe they have deteriorated. In any case, the pilot should know his/her limitations and those of the aircraft to operate it in a safe, sensible manner.

Let's return to square one, the *preflight*. The total preflight begins and ends with the one who is responsible for the safe conduct of each flight—the pilot.

In any worthwhile human endeavor, there must be preplanning. Depending on the type of flight to be undertaken, the pilot has many plans to make. The flight could be anything from a nice Sunday afternoon hop around the patch to a transcontinental flight. The planning will vary, but not very much. But the first and foremost consideration should be the pilot. Is he or she really ready? For each and every flight, the pilots have to ask some questions of themselves.

Fatigue

The complexities of operating an aircraft are more fatiguing than many people realize. The constant attention to heading, altitude, airspeed, radio tuning, and communications can tire a person rather quickly. This is especially true if you are a student pilot or a relatively new pilot. The physical duties, when combined with the mental stress of learning to fly, can tire even the most robust person in a hurry. This means you must learn to pace yourself. If you notice you are making random errors that you normally would not make, you are probably reaching the saturation point and should end the lesson. If your instructor does not realize that you are tired, tell him. Very little learning can occur if a person is fatigued, and the dangers of flying in a fatigued state cannot be overemphasized. Remember, fatigue is indiscriminate; no one is immune.

If you know that fatigue is dangerous, then what steps should be taken to see that it does not overtake you? This depends a great deal on the type of situation you're in. As mentioned previously, a student should tell the instructor when he becomes tired. Hopefully, the competent instructor will notice the symptoms in the student and end the lesson.

If you are on a long cross-country flight, there are several ways to cope with fatigue. Play with the radios. Take frequent crosschecks of the instruments. Change the seat position. Change seats. But don't punch on the autopilot and go in the back for a rest. You might have a much longer nap than you planned on—like for eternity. If you are in a small two or four-place aircraft, get out of the left seat and slide over to your right until you are sitting on half of each front seat. Now you can play the oldest coordination game known to aviation. You place your left foot on the right rudder pedal in the left-hand side and your right foot on the left rudder in the right-hand side. Put your right hand on the right yoke and your left hand on the left yoke. Now try to fly straight and level. If this doesn't keep you awake, nothing will.

Physical or mental fatigue can be very deceptive. It can sneak up on anyone at anytime, and it is frequently deadly. A number of years ago, a student graduated from our University Flight Technology Program. He graduated with his Commercial, Instrument, Multi, and CFI certificates. This young man decided he would enter the aviation field as an air taxi/charter pilot. He was an average, or maybe slightly better pilot and he found a job flying charter for a firm in the north. I saw him a few months later and inquired as to how everything was going. He told me he was "flying his butt off" building a lot of twin time but was a little tired of going places at all hours of the day and night. I told him I was glad he was happy, but he could keep that job. I'm very content to be at home at night (no knock to you air taxi/charter pilots; it's just my personal preference).

The next time I heard his name mentioned was when I was told that he had died. It seems he had been flying parts to some factory in Canada and had made several trips in succession starting in the evening and continuing through until daybreak. On the last trip, just as a new day was dawning, he fell asleep. He slumped over the yoke and went nearly straight in. It was a needless tragedy induced by fatigue. He needn't have taken that last load. He could have told someone to forget it. But he didn't. He probably thought he could make it. Or maybe he chose not to recognize the signs of fatigue. Maybe he was afraid he would be fired if he refused that one more flight. Which is worse? Being fired or being dead? As long as I am breathing, no one will ever push me into a life-or-death situation such as taking just one more flight when I am too tired to do the job safely.

I realize the above-mentioned case is extreme, but it happened. And it doesn't always happen to the other guy. Be aware of the fatigue factor. Plan for it. Do something about it. Stay alive.

Every case of pilot fatigue does not necessarily end in disaster. It happens to all of us to some degree every time we fly. Not too many years ago, my boss and I set out on a

Chapter One

beautiful Saturday morning to pick up a new Decathlon we had purchased for our aerobatic program. We left our home base at Lawrenceville, Illinois, in our Cessna 310 and cruised up the length of Illinois across northern Indiana and well up into Michigan. We stopped there in northern lower Michigan only long enough to sign the purchase papers and have a Coke. We then proceeded across Lake Michigan to a place in northern Illinois where the plane was actually located. It was nearly noon by this time, and as usual, the plane we were to pick up wasn't ready.

Finally, at about 2 P.M., I departed for home in our new aerobatic mount. I was fully fresh in the knowledge that I had at least two hours to play with the new plane during the 225-mile flight home. And play I did.

I leveled at 3,000 feet AGL, tuned the radio to the Pontiac VOR, checked the chart to see that I was clear of any airways, cleared the area for traffic, and upside down we went. Inverted VOR navigation—what fun! Tiring of this, it was time to check all the other little maneuvers that bring such joy. A loop, a slow roll, a snap roll, a hammerhead, an Immelmann (had to reverse my direction from the hammerhead, you know), more inverted flight, and so on. Finally, at about Champaign, I began to come down from the exhilaration of all that fun and games. I looked at my watch and found I had been airborne for over an hour. Champaign was off to my left at least six miles, and according to my route of flight, it should have been directly below me. However, I knew the country well and mentally noted the towns over which I had passed. I tuned in the Lawrenceville VOR, centered the needle, and proceeded home.

As I flew toward home, I noticed a very peculiar feeling beginning to overtake me. My arms felt as though they were made of lead. I couldn't think quite right. The VOR needle kept wandering around (I knew it was just an unreliable radio). Nothing seemed to go right and the last 100 miles seemed to last forever.

As I neared home, a sudden burst of adrenaline hit me and I was overtaken with a case of the "look at me's." Surely I had enough energy left to make at least an inverted pass before landing, I thought. I did—at a safe altitude—and landed.

I taxied the beautiful red, white, and blue machine toward the ramp, eagerly awaiting the accolades from the crowd of people who wanted to get a look at the new plane. As I reached the hangar, I swung the Decathlon around, pulling the mixture as I neared a stop, opened the door, jumped out, and found I could not stand up. My head was swirling and my legs would not support me. My whole body was writhing in agony. Not wanting anyone to know of my plight, I knelt down and pretended to examine the belly of the plane.

I knew that sooner or later I would have to get up or call for help. After a few minutes, I slowly made it to my feet and leaned against the plane. I did my best to try to answer the questions as to how it flew, how it rolled, and so on. My answers were very short.

In the back of my mind, I recalled that I still had to fly our Stearman over to an adjacent field in preparation for an airshow the next day. After an hour or so, I did. I then went home, took two aspirins, and sat down to think.

The things I thought were not very pleasant, and the only reason I have embarked on this self-incriminating expose of stupidity is that I pray you will learn from it. Profit from my mistake by being smarter than I was.

To put it mildly, I was fatigued. I was so tired that I unknowingly put myself in a dangerous situation. All students are taught early in flight training never to exceed their, or the aircraft's, capabilities. I certainly didn't exceed the Decathlon's, but I exceeded my own physical capabilities about as far as I ever care to. Remember that no one is immune from fatigue. No matter how physically fit you are, or how mentally acute, you can still fall prey to fatigue. It plays a role in all types of aviation: instruction, charter, cross-country, acrobatics, and all the rest. So don't begin your flight tired.

Illness

Illness is another thing the pilot must check himself for before each flight. It can range all the way from the common cold to some disease that could incapacitate the pilot. Most diseases of a serious nature, that is, ones that could cause a total disruption of the pilot's functions, should be caught at the first flight physical and would not allow the prospective pilot to hold a current medical. But if you have flown for a number of years, you might have since developed such an illness. (See FIG. 1-1). The type of illnesses I'm talking about are the everyday cold, the flu, and headaches. Everyone experiences these

Fig. 1-1. *Your physical condition can affect flight safety.*

maladies once in a while. They are not generally incapacitating and people tend to suffer silently and ignore them until they go away. If they are smart, they stay home and take care of themselves. But some pilots think it is okay to go ahead and fly with a cold, flu, or whatever. They are, however, putting themselves in a potentially hazardous situation.

Suppose you have a mild case of stomach flu, and maybe you have a flight you feel you just have to make. It should be no big deal, right? Wrong. Did you ever try to do anything constructive while throwing up? You could be on short final and have to try to land your plane. Stay home or have someone else fly you there if you feel you really must go. Suffering from *get-there-itis* has caused more grief than any other form of aviation-related illness.

Medication

If you are sick and taking some sort of medication, should you fly? The FAA has a very short answer for this. *No*. They say you should obtain a clearance from your doctor, preferably an FAA-certified Aviation medical examiner, before you take anything and fly.

Drugs and flying can be a very dangerous combination that should not be taken lightly. However, I believe the FAA is taking it to the extreme in recommending that you not even take an aspirin without medical authorization. My doctor would probably bill me for an office visit if I called and told him I had a slight headache and wondered if he thought it would be all right if I took an aspirin.

Also, it has been my personal experience that I receive a much better explanation of the do's and don'ts concerning drugs and their possible side effects from my pharmacist. This professional is trained in the compounds and formulations and is, in my opinion, much better qualified to inform me of the correct path to take. I'm not saying that doctors don't know their medicines, only that pharmacists probably know them better.

Amphetamines, barbiturates, tranquilizers, laxatives, and some antibiotics are but a few of the drugs the FAA says should be taken only during a non-flying period. They can cause sleepiness, lack of coordination, and reduced reflex action and should be taken only as directed. Remember, the effects of these drugs might last for hours or days after you have finished taking them, so follow the advice of your doctor or you could wind up in trouble.

You should also use good judgment when taking many over-the-counter types of drugs (FIG. 1-2). I once had a doctor (also a jet fighter pilot) tell me he had a very bad experience from taking a well-known over-the-counter antihistamine while flying a jet at 30,000 feet. He related that he was very glad he was not solo at the time. Now he's a doctor and should know when he is apt to get in trouble. Use extreme caution when mixing any sort of drugs and flying.

Here is an example from NTSB File # 3-3883. On December 4, 1976, a 57-year-old private pilot with 2,204 total hours departed the Depere, Wisconsin, airport in a Cessna 150E. The flight was listed as a local solo, noncommercial, pleasure flight. The pilot was observed performing maneuvers at low altitude before spinning into the ground.

Fig. 1-2. *Even most over-the-counter drugs are forbidden for consumption while flying. If in doubt, check with your Aviation Medical Examiner.*

An autopsy revealed .92 Placidyl/100 and salicylate in the pilot's urine. (Placidyl is a barbiturate and can cause extreme drowsiness. Salicylate is more like an aspirin and is often used as a painkiller.) The probable causes listed by the NTSB are physical impairment, failure to obtain/maintain flying speed, and unwarranted low flying. The guy was probably having a good time. Now he's dead. Drugs and flying should be mixed with extreme caution—if at all.

Alcohol

How about alcohol and the pilot? According to one FAA official I talked to, off the record, alcohol is probably a contributing factor in more aircraft accidents than anyone cares to believe. The reason it is so hard to prove is that most alcohol-related accidents are proven through an autopsy. How many nonfatal accidents actually involve alcohol is anybody's guess. If the accident is nonfatal, the FAA might not arrive for hours or days, if they arrive at all. Sometimes they mail out an accident report to the pilot to fill out at his convenience. This allows time for the pilot to make up a good story and swear he hasn't had a drink. Therefore, many of the alcohol-related accidents are pushed to the low end of the spectrum.

Alcohol is a depressant. It slows the reflex action, dulls the senses, and perhaps worst of all, can turn Mike Meakpilot into Ivan Ican. How many pilots, fortified by the remnants of a six-pack, have flown on into deteriorating weather? If you must drink, don't drive. And if you drink and drive, don't drive to the airport.

Here is another example, from NTSB File # 3-3644. On November 6, 1976, a Piper PA-24 departed Evanston, Wyoming, for Salt Lake City, Utah. It carried a pilot and a passenger. The pilot had 488 total hours, 241 in type, and he was not instrument rated. No information is given on the copilot. Near Grantsville, Utah, the plane was observed flying low to the ground. The terrain is listed as containing high obstructions.

The aircraft collided with wires and poles and crashed. Both occupants died. The NTSB lists the probable causes as exercising poor judgment, failing to see and avoid objects or obstructions, and physical impairment due to alcoholic impairment of efficiency and judgment. A subsequent autopsy revealed the pilot had a blood alcohol level of .212 percent and the copilot had .200 percent.

The FAA allows no more than .04 percent alcohol by weight. Anything over this and the FAA considers you as being legally drunk. These pilots were five times that level. They were not able to fly an aircraft with any level of skill and efficiency, but they thought they could and now they are statistics.

Mental Preparedness

Before each flight, the pilot should take an objective look at the state of his mental preparedness. Apathy, anxiety, or any other state of mind that could hinder judgment should be laid to rest before a flight is undertaken. A pilot who has his mind on something other than flying is apt to be heading for trouble.

We all have our good days and our bad, our peaks and valleys of mental readiness. Sometimes the cause is just not known. Pilots are people, and people have problems. It could be trouble with your wife, husband, girlfriend, boyfriend, money, or whatever. My point is that these problems have to be dealt with before or after the flight—not during it.

Apathy is a lack of emotion, feeling, or passion. It is a mental state of indifference, lethargy, and sluggishness. What it really is, in plain language, is a state of just not giving a damn. You could care less if the flight is made, the sun comes up, or anything else. If you attempt a flight in this condition, you could be setting yourself up for a bad experience. You must have your mind clear and ready for the sometimes-difficult tasks of flight. If possible, clear up the problem prior to flight time. Or try to put it *completely* from your mind. If you don't, you might be brought back to reality by doing something like landing with the gear up. That is guaranteed to take your mind off your old troubles and put it to work on a new set.

Apathy fostered by poor instruction or from an inadequate training course can usually be corrected. If you feel your instructor is not preparing himself well enough for your dual flight, get another one. Go to a different school, or sit down and have a heart-to-heart talk with the head of flight training at your present school. These actions can usually clear the air and bring desirable results. Then you can get back to the learning you are paying good money for in the first place.

Student/Instructor Relationship

Let's spend a few moments on the subject of the student/instructor relationship. The learning environment during flight training is a difficult one, at best. Many hours are spent in very close proximity to one other person during which time some very tense moments will occur. The stress of teaching or learning in the closeness of an aircraft cockpit while all sorts of events are unfolding can wear emotions raw. On both sides.

The person has never lived who possessed the ability to get along with everyone in the crowded, stress-filled training cockpit. It is, however, imperative that the student and instructor hold a professional regard for one another, or, at the very least, be able to stand each other. If not, then the natural anxieties produced in any learning situation can be magnified by personality conflicts. When this happens, little learning will occur, and dislikes can become very personal. When this happens, it's time for a change.

Most of the time, the instructor will sense this chill and arrange a trade with another instructor. Then, progress will resume and all will be well. However, some people are so good at hiding their true feelings that one person might not have a clue that the other has a problem. When this is the case, tell him. That's right. Just politely tell the instructor that you are not able to relax with his type of instruction, that you would feel better if you could try it with another instructor. It might be just what the instructor has been praying for!

Anxiety is a much more deeply rooted state of mind. It is a state of mental uneasiness arising from fear or apprehension. The causes of anxiety are much harder to detect, and the cure is usually more difficult than it is with apathy.

A person suffering from anxiety feels that no matter what he or she does, it probably won't be right. It is a self-consuming defeatist attitude and must be overcome before safe flight can take place. Sometimes you must look deeply inward to find the cause of the anxiety. Maybe an open, honest talk with a good friend, flight instructor, or professional psychologist will bring the problem to the fore. Then it can be met head on and conquered. Whatever it takes, it has to be solved before you can perform proficiently as a pilot.

Vertigo

Vertigo, or *spatial disorientation* as it is sometimes called, involves a disorientation in space during which the person involved is unable to sense his attitude accurately with respect to the natural horizon (FIG. 1-3). A pilot suffering from vertigo is unable to perceive by his own senses whether the aircraft is climbing, diving, or turning.

When on the ground, you perceive attitude with respect to the earth by seeing fixed objects around you, by feeling the weight of your body on your feet, and by the vestibular organs in the inner ear. You can orient yourself by any one of these means for short periods of time.

While in flight, however, all three of these normal means of orientation can get obscured or confused (FIG. 1-4). The pilot might only be able to see objects that are in or attached to his aircraft when ground references are obscured by clouds or darkness. Your ability to sense the direction of the earth's gravity, by the weight on your body and

Fig. 1-3. *Unchecked vertigo—look out below!*

through your vestibular organs can be confused by accelerations in different directions caused by centrifugal force and turbulence. For example, the senses are unable to distinguish between the force of gravity and the horizontal force resulting from a steep turn.

Because of this fact, gyroscopic instruments are necessary if you are to fly for more than a few minutes without visual access to outside reference points. The use of such instruments does not ensure freedom from vertigo, for no one is immune to vertigo. They do enable a pilot to overcome it, however, if he trains himself to accept the psychological discomfort that results from acting in accordance with instrument indications and to disregard the false impressions received from the senses. This means *disregard your senses and believe your instruments*! It could save your life.

All student pilots should have the opportunity to experience the sensation of vertigo during their early flight training during maneuvers performed by the instructor. The maneuvers and procedures that can cause vertigo are quite simple. Attempting to read a map or manual while performing coordination exercises or watching the upper wing tip in a prolonged steep turn usually can bring vertigo about. Once experienced and understood, later unanticipated incidents of vertigo can more easily be controlled or overcome. As you gain aeronautical experience, such periods of vertigo will become less frequent, but they can and will persist under instrument flight conditions or at night, so be prepared.

Sometimes closing your eyes for a few moments can help you overcome vertigo, but the best thing to do is to fly according to what you see depicted on the flight instruments and ignoring what you feel.

Remember, pilots are most susceptible to vertigo at night and in any other condition in which the natural horizon is obscured. The best defense against vertigo is to experience it, know you have it, and then fly your instruments. As you will see, even the possession of an instrument rating does not automatically free you from the grasp of vertigo. Although FAA studies have shown that a pilot with an instrument rating is generally better able to overcome vertigo, you are not immune from the effects produced in instrument conditions.

Note this example from NTSB File # 3-3850. On November 28, 1976, a commercial pilot who also held flight instructor and instrument ratings took off with three passengers from the Wilmington, Delaware, airport. They were enroute to Savannah, Georgia. The pilot was briefed by weather bureau personnel that the weather was very low IFR. The pilot filed a flight plan and the four departed for Savannah.

Near Oxford, North Carolina, the pilot began to encounter communication difficulties. Radar showed an erratic flight path, and the controller lost the Cessna 210's transponder reply. Although the aircraft was still airborne at this time, the loss of the transponder and the difficulty in communication suggested a possible electrical failure. The pilot apparently was overcome with vertigo while flying in solid IFR conditions and lost part or all of his electrical system. The aircraft went into an uncontrolled descent, crashed, and killed all on board.

Fig. 1-4. *Vertigo can occur anytime, but you are especially susceptible when the horizon is obscured, such as night or in instrument conditions.*

Chapter One

The weather at the crash site was listed as one-mile visibility with a ceiling of 100 feet, fog, and light rain. Probable cause: vertigo induced by low visibility and ceiling with the possible loss of the electrical system. The point I'm trying to make with this particular report is that even an instrument-rated pilot can run into compound problems.

The loss of the electrical system while on an actual IFR flight is a very frightening thought. But the situation does not have to be hopeless. A well-trained instrument pilot should be able to keep his aircraft under control after experiencing a total electrical failure. (No one knows exactly what happened in the aforementioned accident, so let us go with some generalities. Not having been there, I do not attempt to say what that pilot should or should not have done because the circumstances might have been far more severe than we know.)

In the event of an in-flight emergency of any kind, the pilot's primary responsibility is to maintain control of his aircraft. This is first and foremost. If all electrical systems are lost while on an IFR flight, the pilot has but two choices. One is to panic, experience vertigo, and die. Or the pilot can use his or her training and the remaining instruments, keep as calm as possible, and control the aircraft. There might not be communication or navigation abilities, but these are secondary. You have to keep control of yourself and the aircraft.

Size up the situation. Where are you? Which way to better weather and more suitable terrain? Use the instruments remaining. If the electrical system is gone, it should still leave the compass, airspeed indicator, and maybe the turn coordinator. I realize many of the new turn coordinators are electric, but the gyro instruments are usually run off of the engine-driven vacuum pump, so they should be operative. Use what you have, and maintain control of the aircraft.

Once control is confirmed, maintain it and start the decision-making process that will give you the best chance for survival. You know the forecast weather for the area in which you are flying. Is it better back where you came from, ahead, or in a completely different direction? Remember the terrain. Select the approximate heading that will take you toward the best of both. *Gently* start to turn in the desired direction and don't give up. Check fuses and circuit breakers. Sometimes a blown fuse can be replaced to give communication and navigation for at least a period of time. If so, see how many fuses you have and use them for quick checks. Then remove them. You might even be able to navigate and shoot an approach, call for some aid, or at least warn ATC of your trouble so they can clear the area of known traffic.

Also, battery-powered NAV/COM radios are now available on the market, and such a radio should be in the possession of any pilot while flying in instrument conditions. These little beauties have a battery life of several hours, can provide you with VOR navigation, and let you talk to ATC at the same time. They are a very affordable backup system capable of providing lifesaving navigation as well as communications with ATC.

If you are not able to remedy the problem, you still have a good chance if you don't panic. Keep control of yourself and the aircraft. Turn toward the best weather and terrain. When the time comes to start down, if it's still solid IFR, slow the aircraft to a safe, slow, descent speed and descend with wings level until you either break out or land. If you have set the aircraft up in landing configuration and control the airspeed with pitch and your altitude with approach power, you just might land in a nice big pasture. If not, you will impact the ground under control with wings level, and as slowly as possible. You can do no more.

More likely, when you break out of the clouds you can circle around and find a suitable place to land, maybe even navigate to an airport and put it on a real runway. It can be done. It has been done. You can do it if you don't panic and let vertigo disorient you.

Pilot Licenses

Aside from the physical and mental preparation, there are a few more things a good pilot must check before each flight. Are your medical certificate and license in your possession? Are they current? Are you current in the aircraft you are going to fly this day? If these and all the other things in the previous pages are in good shape, then you are probably physically, mentally, and emotionally ready for the flight. This is as it should be. Remember, the pilot-in-command is responsible for the safe conduct of each and every flight. It's your life and your flight. Prepare well for it and it will probably go as planned. (FIG. 1-5.)

Fig. 1-5. *In order to fly legally you must have on your person your current medical certificate and pilot's license.*

THE AIRCRAFT

After making the personal evaluation of your readiness for a flight, the next thing on the agenda is a thorough check of the aircraft. As far as I am concerned, anyone who fails to faithfully complete a preflight of his or her aircraft is neglecting one-half of the equation. If the pilot and aircraft are both in excellent shape, they will usually overcome problems with weather or anything else that might come along.

However, many pilots believe a total preflight of an aircraft consists of a thorough walk-around looking for any loose nuts and bolts. This is but a portion of a really good preflight.

Logbooks

The beginning of a really thorough preflight starts with the paperwork. As I just stated, the pilot-in-command is solely responsible for the safe conduct of each and every flight. This must include more than making sure the aircraft has fuel, oil, and air in the tires.

I start with the logbooks. The FAA requires that every aircraft owner maintain a current set of airframe and engine logbooks. These logbooks do not have to be carried in the aircraft, but must be readily available. They can be just about anywhere, but you had better know where they are—always.

So what should be evident in these logbooks? What do you look for in order to prove your aircraft was airworthy, at least at the time of the logbook entry? It varies as to whether the aircraft is a privately owned, noncommercial used for your own business or pleasure, or whether it is used for commercial purposes.

If the aircraft is your own and you do not rent, lease, or otherwise make money directly from it, then the first item to check is whether the aircraft is within the time constraints of its last annual inspection. Every aircraft must have an annual inspection, and the only legal place to find this information is in the aircraft airframe and engine logbooks. Look in the logbooks for the latest date at which an annual inspection was performed and the signature of the mechanic who did the work. And be sure the log entry states that the aircraft has met standards for its annual inspection. Now, double-check today's date to be sure you are within the time allotted for the annual.

How long is an annual inspection good for? Sounds like an easy one, doesn't it? An annual inspection is valid for 12 calendar months, after which if you were to fly this aircraft, you would be technically illegal. Your aircraft would be considered unairworthy. But, how long is 12 calendar months anyway? Many pilots get mixed up on this point. And it is easy to become a little confused, so let's look at an example.

If the logbooks state that your aircraft has last had an annual on June 18, 1998, then that particular aircraft would be due for an annual inspection during June of 1999, but it would be legal until 11:59:59 P.M. of June 30, 1999. After that, the aircraft would become illegal for flight. In short, the annual inspection is valid from the date of inspection for one year *plus* the number of days left until 11:59:59 on the last day of the month in which it was inspected. If your aircraft was inspected on the last day of a given month, the annual inspection is valid until midnight of the last day of that month the following year.

All aircraft must have an annual inspection unless they are on a progressive maintenance program (this program is maintained on so few light aircraft that I do not pursue it further). In addition to the annual inspection, aircraft used for hire must have an inspection every 100 flight hours that should be entered in the airframe and engine logbooks. You should look in the logbooks for an entry certifying that the aircraft you are going to fly has had its 100-hour inspection and be sure that it is not overdue for another inspection. Again, if it were overdue, you would be flying an illegal aircraft.

While looking through the logbooks, it doesn't hurt to scan the pages for any sign of substantial damage, how and when it was repaired, and any other pertinent information you can find. You might gain a clue as to where to look more closely in the preflight. Remember, annual and 100-hour inspections do not guarantee that the aircraft is airworthy. It is *legally*, but perhaps a bird has built a nest in the prop governor, or someone backed a fuel truck into it during the night.

Speaking of birds, I once started to rent an aircraft from an outfit of which I had heard bad reports. While inspecting the logbooks, I found a notation about some damage and repair from a bird strike. I inquired as to the significance of this occurrence and was informed it was "no big deal." Upon walking out to the aircraft, I remember thinking that the bird must have been struck while still in a tree. I declined the rental.

Now that you have scoured the logbooks and are sure the aircraft is legal as far as the annual and 100-hour inspections are concerned, what's next? Find the key and clipboard and look to see if there is a *squawk sheet* attached. A squawk sheet is a form used to indicate any malfunctions, gripes, or complaints made by prior pilots after a flight. If one is not attached, ask someone if his or her place of business keeps such a record. If they do, ask politely to cast your eyes upon it. Note the date and type of comments made by the previous pilots about the aircraft you are about to fly. If the sheet is relatively clean and the remarks are only that the number two radio is intermittent or that the seat is uncomfortable, don't worry too much. On the other hand, if you see scrawled across the sheet something like "junk," you might think the situation over a little further.

If you are flying at an FBO or flight school with which you are well acquainted, you might try to find someone who has recently flown the aircraft. Ask if they noticed any particular problems with the aircraft. You can pick up some very useful information this way. But don't take anyone's word that the aircraft is in top condition. Anyone can overlook something very important.

After a thorough check of the airframe and engine logbooks, squawk sheets, and any pertinent information you obtained from other pilots, head out toward the aircraft—don't forget the key! I couldn't count the number of times I have kicked pebbles on the ramp while waiting for a student to go back and get the bloody key. (Of course, I have never forgotten one.)

General Overall Look

As you approach the aircraft, give it a good general overall look (FIG. 1-6). Sometimes you can spot a wrinkle in the skin or some other problem better if you are a short distance away and not right on top of the aircraft.

Fig. 1-6. *As you approach the aircraft, look for any obvious signs that might render it un-airworthy.*

We once had a Cessna 150 whose prop had been removed and sent in for some reason or another. Since the plane was taking up room in the shop, we decided to take it out and tie it down on the line with all the other aircraft. We had fun with that plane. We'd find a person who was unaware of the problem and ask them to look out and tell us what was wrong with the 150. It took some people quite a while before it dawned on them just exactly what the problem was. One instructor thought it might be interesting to take the whole thing one step further. He sent one of his students out to preflight the propless 150.

Off he went. He got to the aircraft, checked the paperwork, drained the sumps, checked the oil, ran the flaps down, checked all the controls, and returned to report the aircraft ready for flight. His instructor nearly died. I didn't relate this little tale to make the student look foolish. There is a message here: DO NOT TAKE ANYONE'S WORD THAT YOUR PLANE IS READY FOR FLIGHT. CHECK IT YOURSELF! As I said, the pilot-in-command is solely responsible for the safe operation of the aircraft, and that most certainly includes the preflight.

There are any number of ways to go about the preflight. It can vary from plane to plane and from pilot to pilot. The main thing to remember is to do it the same way every time. It is probably best to use a written checklist so the procedure never varies.

I teach my students to arrive at the aircraft with their eyes wide open. They have given the aircraft a thorough overview as they approached, so they will know if there is anything major wrong when they arrive at the tie-down.

When you arrive at the tie-down, leave it just like that—tied down. You aren't going anywhere for a while, and it is much safer to leave the aircraft tied down so someone doesn't come by in their King Air, helicopter, or whatever and blow you away. This problem is especially true if you live in any part of the country where the wind is predictably unpredictable. For example, in west Texas the wind can go from 0 to 50 MPH in about the time it takes you to untie your aircraft. This can lead to some interesting square dancing between you and the aircraft. I know. I have participated in this experience several times.

Aircraft Documents

Now it's time to start your thorough preflight. Open the door and take a good long look around the interior. If there is anything significantly wrong, it should catch the eye of an observant pilot. If all seems to be in order, the next step is to check the paperwork. Make sure the airworthiness certificate, registration, operation limitations, radio station license, and weight and balance data, and all placards are in order. Remember, the FAA requires that the airworthiness certificate be in full view at all times. Don't shove it in with the spare fuses or cram it behind the seat with the luggage tie-down. Keep it in *front* of the little plastic window that most aircraft have for document storage.

Here is another area with which many student pilots seem to be unfamiliar. How do you know the airworthiness certificate is valid? Is it valid just because it's there? Is there a date of the last annual or 100-hour inspection imprinted on it to show validity? Is it valid forever? No!

If, during your preflight inspection, you find no obvious damage that would render the aircraft unairworthy, the only other thing that validates an airworthiness certificate is an inspection of the airframe and engine logbooks to make sure the aircraft has had all pertinent AD notes, inspections, and damage repair endorsed by the proper authority. So, providing you have checked the logs prior to arriving at the aircraft and you don't discover anything during your preflight inspection to prove otherwise, then the airworthiness certificate is indeed valid.

The airworthiness certificate has a date attached to it—the date of issuance. However, it doesn't mean there has never been a problem. It only means that on the date of issuance, the aircraft was indeed airworthy.

After a detailed examination of all required aircraft documents, return them to their proper place and continue with the preflight (FIG. 1-7).

Interior Preflight

Next, position the fuel selector so that the engine would draw fuel from a full tank. This may be left tank, right tank, or both in an aircraft equipped with multiple positions, or to the ON position in an aircraft that has only the ON/OFF positions. The reason for turning on the fuel at this point in the preflight is that most aircraft are arranged so that their sumps will not drain without the fuel selector in the ON position. And even if they do drain, you would only be draining the line and not the tank. Any water or sediment remaining in the tank would be settled on the bottom, waiting to flow into the fuel lines when you turned the fuel selector on.

Fig. 1-7. *The interior preflight consists of checking the paperwork, fuel quantity, fuel selector, mixture, flaps down, throttle idle, master, and mags OFF.*

With the fuel selector in a supply position (ON), you should make sure the mixture is pulled full lean, throttle idle, and carburetor heat or alternate air in the cold, or off, position. Make sure the magnetos are OFF. If your aircraft is equipped with a key, leave it in your pocket or place it atop the panel until you are ready to start the aircraft. This method is the most effective way of not accidentally having a hot prop during the exterior preflight. If the key isn't in place, it is probably safe to handle the prop, but never trust it completely. Always treat a prop as if it were going to bite you, because it can.

Next, turn on the master switch and check the fuel gauges. This will give you an indication of available fuel. Don't rely on the indications as gospel, however, as electric fuel gauges can sometimes give false information. When you are doing the preflight, always open the fuel tanks and look in to verify the readings given by the interior gauges.

If your aircraft is equipped with electric flaps, lower them while the master switch is still on. They will then be in a better position to be checked during the exterior preflight.

As you are checking the fuel indications and lowering the flaps, listen for the whirling sound of the electric gyro instruments as they begin to wind up. You will know they are at least trying to function correctly, but they should still be checked in more detail later, during the pretakeoff run-up.

Next, turn off the master switch in order to save battery power, and set all trim tabs for takeoff. If possible, watch for the movement of the trim tabs as you turn the trim tab in the cockpit. This will tell you that they are indeed in good working order and should function properly in flight.

The next item on the agenda will be to remove the control lock. You cannot tell if the control surfaces are functioning properly while the control lock is installed. Unless there is a strong wind blowing, remove it.

Now, go back through and double check to see that the fuel is on, the mixture lean, carburetor heat cold, throttle idle, trim tabs set, magnetos off (leave the key out), and control lock removed. Stow any baggage, maps, etc., and you will be ready to begin the exterior preflight.

Exterior Preflight

As I said before, there are any number of ways to complete the preflight. Do it any way you care to, but *do it the same way every time.* It's probably preferable to use a written checklist and follow it step-by-step. Then you are sure of checking everything in the proper order.

If I am preflighting a multiengine aircraft, I usually start at the door and continue around until I am back at the door, ready to get in and go. In most single-engine trainers, I start at the engine access door and continue around until I arrive back at the starting point. The exterior preflight of a Cessna 152, for example, would go like this: Start at the engine access door, open it up, and take a good look inside. Look for any obvious discrepancies, such as a magneto lying in the bottom of the cowling. Look for security of nuts, bolts, wires, and harnesses. If a wire is disconnected, point it out to someone, preferably a mechanic. If the aircraft hasn't been flown for a few days, check for any signs of bird nests or the presence of mice. Birds can build a nest faster than it takes to say it, or so it seems. All those dry twigs and leaves can start a very hot little fire, but it usually doesn't happen until you are in the air and the engine has heated up to a point that causes the nest to ignite.

Check the oil level and add as necessary. Be sure to wipe the stick clean and then recheck the level if the engine is cold. For some reason, cold oil has a tendency to climb up the stick and give a false indication. You can also receive a false indication if the engine has run recently because some of the oil will remain up in the engine for a time. It might be wise to recheck it again when you arrive back at the engine access door after completing the preflight. Put the dipstick back on securely first lest you forget. More than one pilot has created a problem when a loose oil filter allowed oil to be pumped over vast expanses of landscape until it was noticed—usually by the engine getting hot enough to seize.

One more thing on the subject of oil dipsticks. *Do not* lay the dipstick down. Hold it in your hand until you put it back into place. People have laid a dipstick aside and forgotten it entirely. Besides, if you lay it down, someone might step on it or it might get dirty. In the past, whenever I had to put an oil dipstick anywhere other than in my hand, I used to wipe it off and put it in my pocket. After ruining a few pairs of pants, my beloved wife suggested a place, perhaps best not mentioned here, where I might put the dipstick if I thought I might lose it. As a direct result, I now hold it in my hand.

Oil checked and properly secured, the next thing to do is drain the engine fuel sump. Give it time to drain out any unwanted water or debris. Three to five seconds drain time should be enough for most situations. If possible, it is best to try to catch the drained fuel in a clear container. This way you can see any debris or water that might be contained in the fuel. If any unwanted material is found, continue draining the sump until the fuel is clear. Remember, water is heavier than fuel, so it will quickly settle to the bottom.

If you don't have a glass container to catch the fuel in, draining it out onto the ground and then looking for any bubbles of water also works, only not as well. You can't tell if the fuel contains any debris. (I am assuming you are draining the sump onto a hard surface. If you drain it onto grass, there is no way to tell if any foreign substance is contained in the fuel sample.) Be sure the valve is completely off and not leaking.

Close the engine access door and move toward the front of the aircraft. Examine the cowling for any sign of loose rivets, etc., as you make your way to the prop area. Once in front of the aircraft, look inside the cowl on top of the cylinders for any signs of bird nests or other material that might have collected on them. Be sure all visible wires and harnesses are in place and secure.

Next, grasp the propeller about half way out and gently push and pull it to make sure it is secure. If it comes out in your hands, you can be fairly certain that something is amiss. There should be a *small* amount of movement in and out, but nothing monumental. Check the leading edges of the propeller for any signs of nicks or cracks. As a propeller rotates, it creates a tornadolike suction that can pull up rocks, dirt, or other foreign material. These particles can strike the leading edge of the prop and cause nicks (FIG. 1-8). Even if the nick isn't very large, it is capable of posing a potential problem. Even the smallest nick can weaken the prop and cause it to break off at high RPM. Losing the tip of the prop causes an immediate imbalance that can cause a tremendous vibration. This vibration usually results in a loss of your engine—it doesn't just quit, it can fall off the aircraft. This sets up an almost impossible CG situation, among other things. The tail becomes much heavier than the nose and tends to pitch down as the nose pitches up. There is usually insufficient elevator to pitch the aircraft to level flight attitude and the results are often disastrous. Remember, all of this was caused by a little nick on the leading edge of a propeller. Little things can be very important.

On the preflight, if you find a nick on the leading edge of the prop, take it to a mechanic and have him file it down so there will be no danger of losing the tip. Remember, the prop is a very delicately balanced airfoil. If you continue to nick the prop or sustain quite a number of blemishes on one particular side, it would be wise to remove the prop and send it to a prop repair shop for rebalancing or possible replacement.

Next, squat down directly in front of the aircraft and check the air filter that is located directly beneath the prop and landing light. It should be clean and secure. While in this position and directly in front of the aircraft, check for security of the exhaust pipes and nose gear. I recommend you check the exhaust pipes for security with the toe of your shoe. This prevents you from turning your fingertips to charcoal if the aircraft has run recently.

Check the nose gear for proper inflation, tread wear, and any signs of cords showing through. If the cord is evident, it's time for a tire change. Also check the nose gear for stability and any sign of missing cotter pins, washers, and nuts. I tell my students that while it is important to check the things that are there, it is even more important to find anything that is not there during the entire preflight.

Fig. 1-8. *Always check the leading edge of the prop for any signs of nicks or cracks.*

Fig. 1-9. *The leading edge of the wing should be checked for any obvious problems such as dents, loose rivets, blocked pitot tube, etc.*

With the front of the aircraft secure, move on to the cowl on the opposite side from the engine access door. Check for any loose or missing rivets and be sure the static port, which is located just in front of the pilot-side door, is clean and free of any wax, dirt, or insects. For some reason, insects like to nest in the tiny static ports of aircraft.

Now, to check the fuel, grasp the handhold located on the fuselage and place your left foot on the step provided along the lower fuselage. Step up and place your right foot gently on the step provided in the middle of the wing strut. It is important that you keep the majority of your weight on the fuselage step because the strut is not formed to take much weight at a right angle. If your aircraft is not equipped with steps, it is best to use a ladder. Look in to see the fuel level. If you cannot see into the tank, you can try to feel the fuel level with your fingers or use a *clean* stick to measure the fuel level. As I said before, don't trust electric fuel gauges; you'll feel much better having seen exactly how much fuel you have.

Secure the fuel cap and begin the check of the leading edge of the wing (FIG. 1-9). As you move along the wing, check for any dents, loose rivets, or tears if your aircraft is fabric covered. Also be sure the strut is secure and free from any bends, dents, etc. As you look over the wing and strut, lean down and take an overall look at the wing bottom. Be sure it is free from any problems that might hinder a safe flight.

Moving down the wing, you come to the pitot tube at about the point where the strut joins the spar. It must be free of any foreign matter if it is to supply adequate ram air pressure to the pitot instruments. A reminder: *Do not* blow into the pitot tube to free any debris. This usually results in ruining the airspeed indicator. The force of the human lung has such power that a short blast of air expelled from the mouth can have more pressure than the pitot tube experiences in flight. If you happen to encounter foreign material in the pitot tube, try to pick it out with a paper clip or some other small pointed object. If this fails, you might try blowing *gently* into the pitot tube while someone else watches the airspeed indicator to make sure you don't blow it away. If you aren't sure of what to do, take it to a mechanic and let him solve the problem.

Finish the preflight of the leading edge of the wing and arrive at the wingtip and examine the navigation light to be sure it is secure. While at the wingtip, check the general condition to be sure it has suffered no *hangar rash*. (Hangar rash is a common disease caused by the careless movement of aircraft in or around other aircraft that can result in wrinkled wingtips.)

Moving to the aft section of the wing, pivot the aileron up and down, checking for freedom of travel. At the same time, watch the aileron on the opposite wing to make sure it is moving in the reverse direction. Besides freedom of travel, you want to check the aileron hinges for security and smooth operation, and to be sure that all cotter pins are in place. When examining the aileron hinges and actuating rods, hold the aileron firmly in the up position so that any wind gust will not crush your fingers (FIG. 1-10). Fingers have been lost this way, especially in metal-winged aircraft.

Since the flaps are already in the down position, examine them and their tracks, operating rods, and hinges very carefully. Most flaps will have a slight amount of play in the down position, but not much—maybe a half inch. If there is much more play than that, have them checked by a mechanic before attempting flight.

Fig. 1-10. *When you check the actuating rods of the ailerons, hold the aileron in the up position so a sudden gust of wind doesn't shorten your finger.*

Chapter One

While you are under the wing checking the flaps, it is a perfect time to turn and drain the wing fuel sump. It's within easy reach at this time. As with any fuel sump, try to drain it into a clear container for better viewing of possible contaminants.

While on the subject of checking the fuel, always be certain the fuel you drain out is the correct octane for your particular aircraft. You can tell the octane rating of the fuel by the color. 80-octane fuel is red. 100-octane fuel is green. 100 low-lead fuel is blue, and jet fuel is clear.

There is one very important rule to remember when dealing with aviation fuel. When two different octane grades are mixed in your tank, they will turn clear like water or jet fuel. This is a chemical process that is formulated in at the fuel refinery as a safety measure. It is intended to get our attention when we sump a fuel tank and see a clear liquid.

So, when you sump your aircraft and find a vial of clear liquid, you have three possibilities; you have either a mixture of two different grades (octane) of fuel, water, or jet fuel. Since we know our little grasshopper won't run on either water or jet fuel, and a mixture of two different fuel octanes may or may not be a healthy concoction to feed the aircraft, what should be done?

First, you must identify whether or not you may have a tank of water or jet fuel. Jet fuel has a very distinctive smell like kerosene. If this is the case, you will most likely have to drain the tank and replace it with the proper octane.

If you have a load of water, you should attempt to drain it out through the sump until you get the proper color fuel for your particular aircraft.

After eliminating the water or jet fuel possibilities, you must come to the conclusion that you have somehow received a mixture of two differing grades of fuel. Now the problem is to track down and find out just which two you have in your tank. If the octane that has been placed in your tank by accident is of a higher number than your aircraft requires, then it should be okay to fly. But, if the octane is of a lower number that you require, you may want to drain and replace the entire tank with fuel of the desired octane.

Directly below the fuel sump, check the landing gear, tire, and brake assembly. Watch for any sign of moisture on the ground beneath the inner portion of the wheel. If your brake line has developed a leak, this is one of the primary spots to find an accumulation of fluid. Also check the brake and tire for any signs of unusual wear.

As you move back along the side of the aircraft toward the tail section, check the side and underneath for any loose rivets, tears, dents, etc. Be sure the surface has not buckled from any undue stress or strain.

Reaching the tail section, check the leading edge of the horizontal stabilizer as you did the leading edge of the wing. It should be free of any large dents (although you are apt to find quite a few rock chips and nicks) and fairly solid as you try to move it gently up and down.

The elevator should move very freely and should travel all the way to the stops that are found just beneath the rudder. Now is a good time to check the external rudder cables for proper tension and security. Watch for any loose or missing cotter pins, nuts, or bolts.

Move the rudder from side to side and feel for smooth travel all the way to its stops. Look up and check your VOR antenna and anticollision light (strobe or rotating beacon). If you can reach them, feel for the proper security.

Now, complete the right side of the aircraft as per the methods utilized for the left side. With the completion of the walk-around preflight, you have the peace of mind that comes from knowing you have done everything in your power to determine that the aircraft is ready for flight. It is a comforting feeling.

MISTAKES

During the preflight, many pilots are not as careful as they should be, and some are just plain stupid. Many needless accidents are directly attributable to a careless or nonexistent preflight.

Consider this accident that occurred several years ago to a pilot I knew quite well. I will warn you: This accident is so unbelievable that you might think I made it up. Alas, I did not.

The pilot, who held an Airline Transport Certificate, without benefit of a preflight, hopped into a friend's Cessna 182 with the full intent of flying it and found it would not start due to a dead battery. He got out to turn the engine over with the prop a few times to "loosen the oil." Exactly just what he thought this would do for a dead battery escapes me, but that's what he said.

On the second pull on the prop, the engine roared to life, rushed past this poor boob, and was off on its own. Still on the ground, the aircraft gathered speed as it crossed the active runway, lost its tail as it passed under (almost) an irrigation system, and crashed a half a mile away in a gravel pit. The aircraft was a total loss, as were the nerves of the pilot-in-command, who quit chasing the aircraft after the first 100 yards. (Just a minute. Don't you have to be *in* the aircraft in order to be pilot-in-command? I guess that's a matter for the FAA.)

Consider the number of mistakes this pilot made. I can count a minimum of six options open to this pilot, any one of which would have most certainly prevented this accident. Six simple items that a student pilot would do automatically were completely disdained by this pilot for reasons known only to him.

Options available to him in a random order were: set the parking brakes; chock the aircraft; tie it down; turn off the mags; turn off the fuel; or leave the mixture full lean.

Here we have an ATP-rated pilot, all alone, turning the prop through with the aircraft's parking brakes not set, untied, not chocked down, fuel on, mags on, and mixture at full rich. It boggles the mind.

A $150,000 aircraft is smashed into 500 pieces because an ATP-rated pilot does not have the common sense to use the brain God gave him. Makes a good case for remedial flight training on a yearly basis, doesn't it? The trouble is, although throughout my career I have taught hundreds of students to fly, I have never been able to teach one to think better than they could when I found them. Differently, maybe, but not better.

There are more than enough problems associated with flight without going out of your way to tempt fate. A safe flight begins with proper preplanning and continues from there. Don't shift the odds by a poor or nonexistent preflight of yourself or your aircraft.

Chapter One

THE WEATHER

Each flight must also be viewed with regard to the prevailing weather conditions to be encountered during the flight. Whether the flight is to be a local pleasure flight, dual instruction, or an extended cross-country, the pilot must determine whether the weather conditions are favorable, marginal, or poor. This determination must be an integral part of the total preflight.

Far too many flights end up as statistics due to improper respect for weather conditions. These statistics include instrument as well as noninstrument-rated pilots. The NTSB and FAA files are full of accident reports of pilots who continued flying beyond experience/capability limits into known adverse weather conditions. Some pilots just continue to push on, hoping for the best.

The word *limitations*, so often found in accident reports, is the key to much of the problem. Each and every aircraft has its limitations. Each has its own gross weight, V_x, V_y, etc. Each pilot also has his or her own set of limitations, whether they are known or unknown. You should find your own limitations for a given condition and then never try to push yourself beyond them. These limitations should be flexible. As you gain experience and insight, you might be able to narrow the tolerances you had set for yourself.

These limitations include such items as ceiling and visibility minimums, wind velocity and direction, and duration of flight. Of course, these will vary depending on whether the flight is to be VFR or IFR, local or cross-country, dual or solo. My own personal minimum preference is 500-foot ceiling, 1 mile visibility, about 30 knots of wind at the surface, and not more than 2 hours per leg if on a cross-country. And I avoid ice, unless it's in a glass. I know a few pilots who have purposely flown into known icing conditions just to see how much they could accumulate. I think they are idiots and have told them so. I have had occasions to be involved with ice and have flown my share of 200-and-1/2 instrument approaches, but not if I can help it. I go to my alternate, seek a different altitude, or do just about anything possible to avoid such weather. I have seen too many of my companions die needlessly in bad weather. I have been an instrument instructor since 1968 and have developed a very healthy respect for the forces of weather. Weather is undoubtedly the most powerful force on the face of the earth. There have been times when I have been inside my solid brick home and felt some form of apprehension from a severe storm. It's at these times that I'm especially happy that I'm not aloft in a 3,000-pound aluminum aircraft. Or any aircraft, for that matter. Weather has brought down everything man has ever put into the air at one time or another. Respect it.

If your proposed flight is going to be local, you can usually look out the window and tell if the conditions are such that they favor the safe completion of the flight. But the calm wind you are experiencing at the present time can change drastically in the near future. Possibly the seemingly harmless, broken layer of cumulus clouds is going to gather into a full-blown thunderstorm. There are many possibilities of weather changes in a short period of time. Rather than chance it, why not place a call to the nearest Flight Service Station or National Weather Service?

The friendly people at the FSS and NWS can inform you of present conditions, weather at nearby areas, and what you can reasonably expect to take place for the next 24 hours. They are paid to have such information. You and I pay them with tax money, so why not take advantage of the "free" service? You might even have a facility located on your home base. If so, drop in and get acquainted with the personnel. They will be happy to show you how to read the different charts, give you the forecast, and help you in any way they can. Their job is to serve you.

The advent of the personal computer has brought with it an entirely different group of weather briefings from many varied sources. A pilot can dial up the Weather Channel or AccuWeather to name but two, and gather weather forecasts, winds aloft, etc., for the entire nation and most of the world. These services are usually free, are really quite informative, and are in full color ready for you to print off and take with you on your flight.

Another source which requires a computer and telephone modem at your disposal is the government-sponsored DUAT (direct user access terminal). DUAT will bring you up-to-the-minute weather to the screen of your computer whether you are at home, in your office, or on the road. Although the service is free, you have to decode it and are solely responsible for your interpretation of the weather briefing. I must admit, I sometimes cannot decipher some of the codes without looking them up. A few entrepreneur types are selling software that decode the weather into real words. All this gets into a gray area of whether a pilot can prove he got a weather briefing or not. I mean, there's not a real good way of proving a computer gave you a briefing unless you print it and file it away as proof. I prefer to call the friendly personnel at FSS and let them read it to me. Plus, they keep a log of all briefings given. I like that.

Of course, everyone knows that the prediction of weather is far from being an exact science. Sometimes they are wrong. One thing I have found over the years is they are far more accurate in predicting bad weather. Usually, if they tell you it's going to be bad, it is. If they tell you it's going to be good, it generally doesn't get too bad. They don't miss by much, but if you are going to be totally prepared, you must exercise common sense and compare your own limitations to the forecast.

If you are embarking on a cross-country flight, the same rules apply, only more so. You have no way of *really* knowing the weather in the next county, let alone the next state. So don't rush off into the sky just because the weather looks good. It can change very quickly. Get a report from the nearest FSS or NWS, and get the odds on your side. Many pilots who don't, never live to regret it. On the other hand, there are many pilots who utilize the service for a weather briefing and then attempt a flight regardless of the forecast and end up in trouble.

NTSB File # 3-3803 is an example. On December 31, 1976, a 59-year-old, non-instrument-rated private pilot took off from the Auburn Municipal Airport in Auburn, Washington. The intended destination was Royal City, Washington. Witnesses reported seeing the aircraft take off, enter an overcast layer in a climbing turn, dive steeply, pull up into the clouds again, and then dive into the ground. The weather at the site was 150-foot ceiling with one mile or less visibility, with fog. The pilot had been briefed by FSS personnel, and the forecast was substantially correct. The type of weather was IFR. Type of flight plan—none.

Why a noninstrument-rated private pilot would deliberately take off into a 150-foot ceiling with one-mile visibility is beyond my power of reason. Yet, it's done over and over again. Reread this accident report. Think it over. There are many lessons we can learn from this one very short flight.

How does one go about obtaining a thorough weather briefing? It's really very simple and only takes a few minutes. Let's say you are calling the nearest FSS by phone. Here are some guidelines that will help both you and the FSS personnel. Don't just call and ask, "What's happening in Chicago?" To FSS personnel, this is the first tip-off that they are talking to a nonprofessional pilot. You must be specific and clear and know what you want. Open the conversation by telling them who you are, where you are, and where you are going. Then, tell them when you propose to leave and if the flight is to be VFR or IFR. Also include your proposed route of flight and aircraft type (FIG. 1-11).

Now they have the appropriate information and will inform you of all pertinent weather data, such as general weather along the route of flight, winds aloft, terminal forecasts, area forecasts, NOTAMs, etc. Be sure to write down the information, as you will have a very hard time remembering which city had what if the forecast is anything but beautiful. If you are unclear on any point, or if they happen to leave out something you need, ask for it. Many times an airport has a NOTAM and they forget to inform you of it. I always ask for any pertinent NOTAMs. It beats getting ready, flying to an airport, and finding it closed for repairs.

Fig. 1-11. *Many pilots now use personal computers to access current weather information. It is quick and efficient.*

A pilot, whether a student or more advanced, is going to have to deal with some form of weather from the time he unties the aircraft until it is safely tied back down. Get to know the weather and the people who predict it. Set your own limitations and stick to them. You will be glad you did.

DRUG TESTING

The FAA, in a move questioned by many pilots, offered its solution to the world's drug problem with Advisory Circular # 121-30, issued March 16, 1989. It states, in part:

> "On November 21, 1988, the FAA published its final rule in the Federal Register with an effective date of December 21, 1988. This rule prohibits an aviation employee from performing a sensitive or security-related function if that employee has used drugs as evidenced by a urine test showing the presence of drugs or drug metabolites. Since this rule is designed to ensure a drug-free workforce, certain segments of the aviation industry are required to establish an anti-drug program."

What does all of this say to a pilot? It says that if you are going to fly for hire, you will be tested for drugs. At the present time, all ATC personnel, airline cockpit crews, flight attendants, dispatchers, maintenance employees, security personnel, air-taxi pilots, crop dusters, banner towers, bird chasers, hot air balloonists who give rides to sightseers, and just about everyone else who makes a living in aviation is subject to drug testing.

The aviation industry, as well as the nation, is deeply divided over the drug testing issue. Most pilots don't have a problem with testing any pilot after an accident or incident as an addition to the fact-finding process. The problem many pilots have is that they object to unilaterally giving up their rights as Americans, protected, up until now, by our Fourth Amendment right to privacy.

Under pressure from Congress, the FAA rushed into the drug testing business without a plan. What ensued was a folly of the fifth magnitude. All pilots were summarily issued a summons to drug testing without evidence of any real problem existing in the first place. For instance, as of this writing, there has been one airline accident attributable to drugs since the dawn of aviation. One! And there has never been an accident during an instructional flight attributable to drugs. Sort of like killing a rabbit with a nuclear missile, isn't it?

Then there's the "safety-sensitive" issue. "Safety-sensitive" brings to mind a different set of standards, depending on where your values lie. As for myself, I would like a clear-headed surgeon, school crossing guard, dentist, autoworker, Congressman, ad infinitum.

Well, it's here, but it might not be here to stay. Perhaps more intelligent thinking will come to the fore, but I doubt it will be soon; so if you are going to fly for hire, be straight, or get straight, and pray a false positive test doesn't ruin your career, as it has already done for some.

THE DECISION AND FLIGHT PLAN

The decision and the flight plan, as I see it, encompass much more than the mere filling out of a piece of paper. Filing a flight plan is the culmination of the preflight process.

After taking into consideration the readiness of yourself, the aircraft, the weather, and any other pertinent data, you have to make the decision, "Do I go or not?" Remember, it's your decision. The pilot-in-command has the responsibility for the safe conduct of each and every flight. Don't let anyone influence your decision. Don't let your wife, mother, boss, flight instructor, or anyone else force you into a flight situation you feel is unsafe.

If your decision is to go, then you should file a flight plan (FIG. 1-12). Many pilots only file a flight plan if they intend to go on a cross-country. It's really not a bad idea to file even if you are planning to remain in the local area. With a flight plan, someone knows where you are and approximately when you are to return. If you should run into any trouble, someone will be looking for you about 30 minutes after you are scheduled to close your flight plan. This is cheap insurance since the most it can cost is one short phone call. You can even file your flight plan on your computer by using the DUAT I mentioned previously.

A flight plan is mandatory only if you intend to go IFR. You can go just about anywhere you want VFR without having to file a flight plan. You must admit, it's tempting. You are confident. The day is fresh. The aircraft checks out perfectly. Why file a flight plan? What could possibly go wrong? Almost anything you can imagine, and more!

U.S. DEPARTMENT OF TRANSPORTATION
FEDERAL AVIATION ADMINISTRATION

FLIGHT PLAN

(FAA USE ONLY) ☐ PILOT BRIEFING ☐ VNR ☐ STOPOVER | TIME STARTED | SPECIALIST INITIALS

1. TYPE: VFR / IFR / DVFR
2. AIRCRAFT IDENTIFICATION
3. AIRCRAFT TYPE/ SPECIAL EQUIPMENT
4. TRUE AIRSPEED — KTS
5. DEPARTURE POINT
6. DEPARTURE TIME — PROPOSED (Z) / ACTUAL (Z)
7. CRUISING ALTITUDE
8. ROUTE OF FLIGHT
9. DESTINATION (Name of airport and city)
10. EST. TIME ENROUTE — HOURS / MINUTES
11. REMARKS
12. FUEL ON BOARD — HOURS / MINUTES
13. ALTERNATE AIRPORT(S)
14. PILOT'S NAME, ADDRESS & TELEPHONE NUMBER & AIRCRAFT HOME BASE
15. NUMBER ABOARD
16. COLOR OF AIRCRAFT
17. DESTINATION CONTACT/TELEPHONE (OPTIONAL)

CIVIL AIRCRAFT PILOTS. FAR Part 91 requires you file an IFR flight plan to operate under instrument flight rules in controlled airspace. Failure to file could result in a civil penalty not to exceed $1,000 for each violation (Section 901 of the Federal Aviation Act of 1958, as amended). Filing of a VFR flight plan is recommended as a good operating practice. See also Part 99 for requirements concerning DVFR flight plans.

FAA Form 7233-1 (8-82) CLOSE VFR FLIGHT PLAN WITH __________ FSS ON ARRIVAL

Fig. 1-12. *An FAA flight plan is the cheapest insurance you can have.*

One morning several years ago, I departed my home base at Lawrenceville, Illinois, enroute to Dodge City, Kansas, where I was to speak at a convention. The weather looked pretty good, calling for only a chance of light rain in middle and eastern Missouri, so I elected to take our Cessna 180, which was the fastest single-engine aircraft we had at the time. In taking the 180, I made a trade: speed over reliable radio equipment. It was a trade I would later regret.

I had my course plotted direct to Dodge City and was humming along (with no flight plan) thinking about the speech I was to make. As I neared St. Louis, I noticed a definite increase in the cloud build-ups and turbulence level. The "chance of light rain" was there all right, only in buckets. I thought it must be an isolated shower and turned south to get around it in the direction of supposedly better weather. If anything, it was worse.

I again reversed my course and headed toward a highway that I knew led to the vicinity of Spirit of St. Louis Airport. The rain was coming down very hard by this time and I was caught in the middle of it. Visibility was down to less than a mile and was most definitely IFR weather.

As I neared Spirit, I called and requested an immediate special VFR clearance into the airport. They did not hear the portion of the transmission in which I requested the special VFR. They thought I had requested landing clearance and informed me the field was IFR. I was well aware of that. I again called and requested a special VFR and they again told me, in no uncertain terms, that the field was closed to VFR traffic due to IFR conditions.

By this time, the weather was so bad I knew I had to land soon or I would be forced to climb, with no clearance, to VFR on top and hope I could return to the east to find a hole and get down. I didn't like the thought of the latter. Climbing to VFR on top without a clearance in as busy an area as St. Louis could have placed the lives of many people in jeopardy.

I called them one more time and informed them that they could either allow me to land at their airport or I would put down on the highway just north of the field. (I was really going to land at Spirit no matter what since I had been monitoring them and knew they had little traffic.) So what if I lost my license for 90 days? I figured that would be better then losing my life.

With that, I finally got an immediate special VFR into Spirit and landed uneventfully—at least it was uneventful until I had landed, taxied in, parked, and gotten out of the aircraft.

A man walked up and informed me that I was wanted in the tower. Seems the tower chief wanted to have a few words with me. I had suspected he might.

When I got over to the tower, the boys wanted to know just what I thought I was doing. I explained the circumstances to them and they gradually began to calm down. When I showed them my Commercial, Certified Flight Instructor, and Instrument Instructor Certificates they finally started to relax. Before they saw these certificates, there was talk of jails, hangings, loss of license, etc.

It seems they were afraid they had a student pilot on their hands and were trying to be careful not to upset him (while in the air) by asking certain questions. Upset him? With that, the prosecution crossed over to my corner. I asked them if they thought telling a lost student pilot in IFR conditions to go away was in the best interest of safety. I told them it was the same as telling him to go die somewhere else. Don't do it at our airport.

Once more I asked them why they didn't do everything in their power to see that I found the airport and landed safely? Why, if they thought I was a student pilot, did they just tell me to go away? They had no answer. Silence.

I freely admitted my mistake. I messed up. But I had enough experience to have several alternatives. Would a student? I think not. He would need all the help he could possibly get, such as kind words, reassurance, and headings.

That discussion ended the little incident at St. Louis. I think we all came away a little wiser. The point I'm trying to make with all this is that people are human and humans make mistakes. I do. You do. The people in ATC do. We must all work together to make aviation as safe as possible.

Here is another important point: Don't take anybody's word for anything. The weather forecast for my trip was a joke. Learn weather. Learn clouds and the clues they can give you. I assumed the forecast to be correct. I should have known better. Even as a very experienced pilot, I nearly got into an impossible situation.

Be ready for your flight. Obtain all possible data, ready yourself and your aircraft, and then figure everything will probably be incorrect or incomplete. Usually it won't be, but it's better to be ready if it is. Over the years, I have found that the best pilots are usually the ones who know as much about what *not* to do as what *to* do.

Flying is, and should be, a very enjoyable way to spend time or to travel. And it's safe. But it's only as safe as the pilot-in-command cares to make it. Certainly there are always the unexpected mechanical problems that contribute to an aircraft accident, but far more accidents are induced by human error than by mechanical failure. Statistics bear it out. Rather than bore you with a long list of statistics, please believe a pilot who has made almost every type of aeronautical mistake possible and lived to tell about them only through the grace of God. Be prepared. Start each and every flight as though it might be your last; if you don't, it might be. These ideas comprise the importance of the correct flying attitude.

If you are a little apprehensive as to your abilities in an aircraft or a little unsure as to the correct procedure for a given maneuver, for your own sake and for the sake of others with whom you share the sky, dig into the following chapters on mastering the maneuvers. I do not claim to have all the answers, but I guarantee that if you read through the following chapters, whether you agree or disagree, at least you will think. And a thinking pilot is a safer pilot.

2
Mastering the Maneuvers

A GOOD FRIEND OF MINE HAS WHAT I BELIEVE IS THE PERFECT philosophy of flight. He says, "I don't have to go back to the basics like so many other people preach. I never left them." How true it is. If only student pilots would learn the basics. I mean *really* learn them and then never let them rust, aviation would become much safer.

We all had to learn basic flight techniques. Very simple. They are so simple that many pilots and flight instructors only learn them superficially. They don't really understand why and how a given maneuver works. Maybe they appear to have a good working knowledge, but in reality they are doing the maneuver by rote. Often, this type of maneuver ends up as a statistic. Sound negative? Maybe it is, but so is death and dismemberment. The FAA has long preached the benefits of positive motivation. For the most part, I agree, except I sometimes have difficulty distinguishing between positive and negative. What's more, I sometimes wonder which one is really most effective.

Positive motivation is supposed to create a relaxed atmosphere where learning can be more readily fostered. Fine. Positive motivation is also supposed to show the student that an incorrect response will lead to an incorrect act. Good. But does the other side merit any mention? I feel it does.

Chapter Two

Negative motivation has a very real place in flight instruction. A student must learn the correct procedure in any manner that will result in the ultimate goal—a safe pilot.

The FAA seems to want you to teach a student to fly safely and at the same time sort of play down the dire consequences of an incorrect act. When I'm in an airplane, I'm there for one reason and one reason only: to teach correct flight technique and safety. If I wanted to enter a popularity contest, I'd run for office. I don't mean that you have to be discourteous or unreasonable, but you have to get results whether the student likes you or not. I feel respect is far more important than friendship (incidentally, they usually come in that order: respect and then friendship).

Suppose you were the flight instructor working with a student who was having a degree of difficulty in lowering the nose sufficiently after stalling an aircraft. Which one of the following methods would evoke more awareness and action?

1. "Now, Mr. Smith, if we don't get our nose down a little more on recovery, we are going to be in for a rather big surprise and it will ruin all our fun." Or

2. "If you don't relax your back pressure enough to get that air flowing over the airfoils in sufficient quantity to regain control and lift, you're going to restall the aircraft and kill yourself. Then your wife will collect all that insurance money and run off to Mexico with your insurance agent."

Maybe the second example is a little extreme, but not really that much. To learn, you must gather facts and data and associate and correlate that material into prompt action. It should not be sugarcoated. Facts are what it's all about. Any really good pilot wants to know much more than just *how* to fly. He also wants to know why the plane flies and where he fits into the big picture. Why some people pay good money to learn to fly and then don't really pay attention is beyond me. Maybe the snob appeal has something to do with it. Sure it would be nice to go to the office on Monday morning and tell the gang that you and the wife flew to Guam over the weekend to get a cup of coffee. I have known several people in this category. Most of them are no longer flying. One died after running into some cumulus granitus (mountains) returning from Florida *under* bad weather. Not over, around, or through, but under—in mountains! This needless tragedy could have been avoided by some common sense and basic flight technique.

Horses and airplanes have a great deal in common. Both are very effective modes of transportation. They can take you to just about any place you want to go, day or night, fair weather or foul, but both require a great deal of care if they are going to perform to their potential. They must be inspected, fed, cleaned, and generally cared for so that you know to the best of your ability that they will not break down on you.

Most importantly, the rider, or pilot, must know what he is doing if he is to stay on top of his mount. The key is to stay on top of the situation, not to merely go along for the ride. Do you remember the first time you rode a horse or flew an aircraft? Did you have the feeling, "Where is this thing going next?" We didn't have command of the situation because we weren't sure what was going on. We just went along for the ride and hoped everything would work out all right.

Many people continue along and hope that everything will fall into place of its own accord. Chances are, it won't. You have to work at it if you want to become proficient at a given task. You must study, ask questions, try, fail, and try again. This is learning. It doesn't just happen. It is made to happen by an individual who has the right mental attitude and will not accept anything less than knowing everything about a situation. There are no shortcuts. You have to start at the beginning and build. Remember, there are only two types of people in an aircraft: pilots and passengers. If you fall into the second category and are flying solo, you are bound to be in for an interesting ride.

BASIC CHARACTERISTICS

All airplanes have the same basic characteristics. They all pitch, roll, and yaw (see FIGS. 2-1 and 2-2). The fact that one airplane is 500 times heavier than another is of little significance when you are thinking in terms of why they fly. They are brothers of the air.

The size of a 747 is large by any standards. But does it take a person of superhuman abilities to fly one? Of course not. The captains who fly this large aircraft are just like you and me, but they happen to fly something different from what you and I are presently used to.

Fig. 2-1. *A Cessna Conquest I has higher performance and is slightly more complex, but it's still only an airplane. Given the time, opportunity, and some experience, it is easily mastered.*

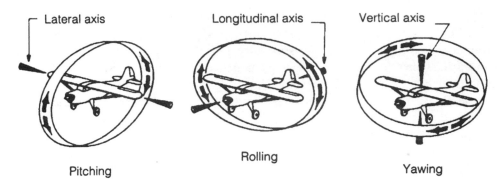

Fig. 2-2. *The three aircraft axes. Roll, pitch, and yaw correspond to lateral, longitudinal, and vertical axes, respectively.*

If you have any doubt that with the proper training and checkout you could fly any aircraft in the world, then you are either lacking in self-confidence or in your basic flight technique. The cockpit in FIG. 2-3 initially looks complicated, but it can be learned as easily as any other. I have absolutely no doubt that I could fly anything with wings, given the proper training. And so could you, if you were properly schooled on basic flight technique.

THE FOUR FUNDAMENTALS

Straight-and-level flight, climbs, glides, and turns are the foundations upon which all normal flight is built. Every single maneuver has its origin in one or more of these fundamentals. Sounds simple, doesn't it? It is, and yet there are many pilots who are significantly deficient in one or more of these basic fundamentals—deficient to the point where they are unsafe to themselves and possibly others as well.

Straight and Level

Straight and level is one of the first maneuvers a student pilot learns. It is the ability to hold a constant heading, altitude, and airspeed. A good definition for straight and level would be slight corrections from any climb, turn, or dive (FIG. 2-4). It is the most fundamental of all the fundamentals. At the same time, it is also one of the most overlooked. Almost everyone believes he or she can fly straight and level very well. But when was the last time you really *practiced* straight-and-level flight? I know what you're thinking. You practice straight-and-level flight all the time, right? You do it out to the practice area and back, and just last week you went on a cross-country. But do you really try to get on an altitude, heading, and airspeed, and nail it? If not, let's explore straight and level in a little more detail and come up with some hints to perfect your straight-and-level flight technique.

Try a little self-diagnosis. Set up straight and level on a given heading, altitude, and airspeed. Note all the numbers. Now, fly for five or 10 minutes and really try to maintain

Fig. 2-3. *This Cessna 421 panel is really only slightly more complicated than a Cessna 152—if you take a close look, the majority of instruments and levers are simply duplicates.*

Fig. 2-4. *Straight and level flight can be thought of in terms of slight corrections for climbs, glides, or turns.*

all those numbers. If the air is relatively smooth, you should have no trouble maintaining near-perfect heading, altitude, and airspeed. Bet it took more work and concentration than you are used to, right? If this is the case, you need to practice much more frequently. Your flight instructor has told you to go out and practice stalls or slow flight, or you have decided to practice these maneuvers on your own, but how often have you been told to, or even thought of, practicing straight-and-level flight?

Straight-and-level flight can be made easier if you use all the elements you have at your disposal. The FAA recommends, and I concur, using the integrated method of flight unless you are in actual instrument conditions. *Integrated flight technique* requires that you use both visual reference and instrument reference. Utilization of this method can do much to improve overall performance because you don't use either reference exclusively. It is a blend of both that aids in improving performance because you are constantly shifting your attention between the instruments in the cockpit and outside visual references. The most immediate gain is the improved ability to spot other traffic. You must always be visually alert because mid-air collisions are *always* a possibility.

In integrated flight, you use everything at your disposal to help keep things where you want them. Glance at your instruments to make sure they are stable at the desired heading, etc. Then shift your eyes outside to get a view of attitude by visual reference. This way, you not only see your attitude in relation to the horizon, but you can confirm what you saw on the instruments. It works wonders for basic flight technique and at the same time improves safety and coordination. Most pilots utilize this method without realizing they are doing it. It becomes a habit and a very good one.

When most pilots look at their instruments, they basically know what they should see. But many pilots have some degree of difficulty in correlating instrument interpretations with outside visual references. When your eyes shift from the instruments for more than a few moments, the result is usually the loss of heading, altitude, etc. The problem is that you cannot always perceive your attitude visually without the use of the instruments; however, the problem is not overwhelming and can be overcome with adequate instruction and practice.

When you look out of the aircraft, whether you realize it or not, you pick reference points and then use them to gauge your position relative to the horizon. The trouble is that sometimes you (unknowingly) pick wrong references. This is natural and should be no great cause for alarm; however, if this habit continues for long, your basic flight technique and all subsequent maneuvers will suffer. You must find correct reference points.

The best instructor in the world cannot tell you exactly where to look, only help you find the best points for reference. Then, it is up to you to take it from there and learn by *trial and error*. The person who never makes a mistake never truly learns. The person who makes the same mistake over and over again is not learning either. The person who really learns is the one who profits from mistakes and then tries another method.

There are several ways to get the feel of straight-and-level flight through the use of outside visual references in combination with instrument interpretation. First, set up straight-and-level flight using mostly instruments. Then, look out over the nose and get a good picture of what you see. Try to take a picture with your mind of what you want to see every time you are straight and level. In other words, you want to see how much ground and horizon are visible and where the horizon line crosses the windshield. You might also note where the cowling of your aircraft is situated with respect to all the other points. If anything changes, then so does your attitude.

For example, if the spot on the windshield where the horizon usually crosses is above it and you have not moved in your seat, you are climbing. Conversely, if the spot is below the horizon, you are diving. It's that simple. And these little aids will be of great help in finding your desired pitch attitude.

Another way to see it is to remember how much ground and sky fill the windshield in straight-and-level flight. If you begin to see more ground than usual, you are nosedown. If you see more sky, you are nose-up. But remember, if you move, so do your references. Don't rope yourself down, but find a comfortable seat position and stay there. Something as small as a change in posture will change your references. Until you have gained enough experience to adjust to changes without hurting your performance, sit still.

Laterally Level Flight

Almost the same procedure is used for *laterally level* (wings level) *flight*. Look out each side of the aircraft and see where the wings are in relation to the horizon. There should be an equal amount of sky-to-horizon distance under each wing in the case of a high-wing aircraft; use top of wing-to-horizon distance above the wing in low-wing aircraft. If not, you are in a *banked attitude*. When you see such a difference, move slight aileron and rudder toward the high wing and the distance will equalize quickly. Then, neutralize the controls and continue your scan of instruments and visual references.

Laterally level flight can also be done quite readily through the front windshield. Simply monitor the horizon. The *cant* (slanted shape) of the aircraft's cowl sometimes makes it appear that you are not laterally level in flight. This causes problems for some pilots, and it might take a little time to adjust to this situation.

Sometimes a pilot subconsciously allows the weight of his hand to pull the yoke down to one side, causing the aircraft to bank in that direction. Holding the yoke lightly alleviates this problem.

The pilot must also hold his head upright to see everything in proper perspective. If the aircraft is in a 5-degree bank to the left and you subconsciously tilt your head 5 degrees to the right, the horizon will appear to be in its proper alignment, when in reality, things are not as they should be. The sooner you learn to ride *with* your aircraft and not to lean against the turn, the sooner you will master attitude flight control.

Learn to properly use the *elevator trim tab*. This little wheel can do much to aid your airspeed and attitude control. Let it work for you and you will be pleased with the results. The elevator trim tab is in there for a good reason. Although many pilots use the trim tab more as a decoration, it is designed to alleviate pressure on the elevator control during all flight attitudes. It is also very easy to use.

Get your aircraft set up in the desired attitude, and rotate the trim tab to "trim off" any pressure you feel. When you no longer feel any pressure on the yoke, let go of everything and the aircraft should remain in that attitude. If it doesn't, put the aircraft back where you want it and trim it a little more. The most important thing to remember when using the trim tab is that it is *not* designed for use as a primary pitch control. It is used to complement your primary pitch control, the elevator, by reducing control pressure in a given flight attitude.

If you are as lazy as I am, you can let the trim tab do a lot of the work for you and at the same time vastly improve airspeed and attitude control. It will work for you in any flight attitude requiring constant pitch.

Climbs

Climbs are another of the four basic fundamentals that must be mastered to achieve co-ordinated flight. Climbs should also be practiced until they are smoothly done using the integrated flight technique.

Many of the same reference points used in straight-and-level flight can be used for climbs. Reference to wingtips can be used to complement instrument indications for laterally level flight and can also be used for pitch information. I teach my students to set up a climb attitude and trim the aircraft for hands-off flight. I then have them look at the wingtips and witness the angle of the wings in relation to the horizon to get a visual picture of the climb attitude. In most trainer aircraft, such as those in FIGS. 2-5 and 2-6, you are usually sitting either under or on top of the wings, so you can get a very good visual reference of pitch.

In most aircraft, you lose the entire forward view of the horizon while climbing. Therefore, the flight instruments become rather important for directional control. However, the wingtips can still be used for directional reference after a bit of practice (FIG. 2-7).

After setting up into a climb attitude, gently apply pressure on one rudder pedal and then the other. Keep watching the wingtips. They will appear to move forward and

Fig. 2-5. *Cessna's 172 is a proven primary trainer.*

Fig. 2-6. *Bellanca's Decathlon is a fine trainer that does something extra—aerobatics!*

Fig. 2-7. *Using the wingtip for reference, this is what you would see from the pilot's seat a Cessna 152 climbing at 80 knots.*

backward as the rudder pressure is applied. If you see them moving, you had a reference point picked out whether you knew it or not. Merely keep the wing on that reference point and you will keep your heading on target. If you keep your wingtip on a solitary point, such as a tree, for any length of time, you will go around in a circle. So a little bit of common sense must accompany this method.

I have found that a distant road, railroad track, or section line is very useful as an aid to this method. But it had better be going in the same direction as you are or you might wind up in Cleveland.

Also, as you continue to master visual and instrument climbs, don't forget to use that trim tab. If you keep yawing to the left, you have probably forgotten to make a correction for torque and P-factor with your right rudder. *Torque* is defined as "any rolling or twisting motion." It is a prime example of Newton's third law of motion, which states, "for every action, there is an equal, but opposite, reaction."

Torque force is present in all propeller-driven aircraft. When a propeller turns, it causes action. Because the propeller is attached to the aircraft, the reaction comes in the form of the aircraft trying to roll, or twist, in the opposite direction. Because propellers on most American-made, single-engine aircraft turn clockwise when viewed from the cockpit, the resultant yaw is to the left. Torque is present from the time the engine begins to rotate until it stops. Its intensity varies only by changing RPM. It isn't noticed in straight-and-level flight because the aircraft is designed so no correction is needed in this attitude.

P-factor, on the other hand, comes into play only during climbs and descents. It tends to yaw the aircraft to the left in a climbing attitude. P-factor yaws the aircraft due to unequal thrust. The aircraft is pulled through the air by the propeller, which is an air-foil. It is literally a rotating wing. As a two-bladed propeller rotates in straight-and-level flight, both blades get an equal "bite" of air, which produces an equal amount of thrust from each blade. However, in a climbing attitude, the *descending* blade (the one on the right when viewed from the cockpit) gets a much larger bite of air and creates more thrust than the *ascending* blade, which is practically loafing. Because of this unequal thrust, the aircraft is yawed to the left. The proper corrective measure to offset the effects of both torque and P-factor is the application of right rudder as needed to maintain heading.

While you are practicing climbs, don't forget to monitor attitude and airspeed control should your attention be diverted from the cockpit. On December 29, 1976, a Bellanca 7ECA stalled and mushed into the ground near Tecumseh, Nebraska. The private pilot had 766 total hours, 290 of which were in this type of aircraft. Ceiling and visibility were unlimited, and there was an 18-knot wind. The remarks section of this accident report sums it up thusly: "Low-level flight in conjunction with a coyote hunt. Stalled on leeward side of trees."

It's a safe bet that the pilot was concentrating on something other than his flying and didn't notice his attitude or pay any attention to the cues received by his senses. He stalled and crashed.

Descents

Descents are the third of the four fundamentals and should be honed to perfection. Too many pilots are inclined to merely push the nose down and let the chips fall where they may. While this might get the job done, it is not the way to obtain maximum performance from your aircraft. Unless you like to see redline speeds on the airspeed indicator and feel the wings vibrate to the point that they might leave you, there are much better ways to accomplish a descent.

In teaching basic descents, I have my students utilize the two most-used types of normal descent procedures. One is the *cruise-type descent*, and the other is the *approach-to-landing descent.*

It doesn't really matter which one is learned first, as long as both are understood and mastered. The cruise descent is used mainly on a cross-country-type flight to reduce the altitude gradually over a relatively long period of time and distance. Like all other aspects of safe flight, it requires preplanning and practice until it becomes a good habit.

Let's assume you are arriving in the vicinity of your destination airport after a cross-country and you have to let down from 5,000 feet to an airport at sea level. If you have found that a comfortable descent rate for your aircraft is 500 feet per minute (FPM), then it will take you about 10 minutes to come down. How far out will you have to begin the descent in order to arrive at the destination airport at the traffic pattern altitude? If the destination airport has a 1,000-foot traffic pattern, you will have to lose 4,000 feet. At 500 FPM, it will take about eight minutes. For simplicity, let's say you have a groundspeed of 120 MPH. Remember, it's groundspeed, not indicated airspeed, that will give you the most accurate information. However, if you do not know your groundspeed, indicated airspeed will probably get you close enough unless you have a very strong wind.

You have eight minutes to lose 4,000 feet at a groundspeed of 120 MPH. Since 120 MPH equates to two miles per minute and you have eight minutes to get down, merely multiply the miles per minute by the minutes of descent in order to find the distance out to initiate your descent. In this case, it would be $2 \times 8 = 16$. You will start your 500-FPM descent 16 miles out in order to arrive at the destination airport at a 1,000-foot traffic pattern altitude.

To execute this descent in a precise manner, you will have to practice in your particular aircraft until you find which combination of power, trim, and airspeed gives you the most satisfactory results. My students learn this procedure in the following manner.

Start at normal cruise speed at a given altitude, and slowly reduce power as you trim the aircraft to maintain the cruise airspeed. Watch the vertical-speed indicator as you begin to descend. When it settles on the desired rate, trim the aircraft to maintain the airspeed and use your power to control the rate of descent. It's that simple. Sit back and monitor other traffic arriving at or departing from the airport.

The other type of descent is even more important to the student pilot. It is the descent at normal approach speed. This descent not only teaches another method of getting down; it also helps you gain the feel of the aircraft in approach configuration and airspeed. It is a more precise maneuver than the cruise descent because in addition to having to adjust power and trim, you also have to put down flaps and gear, all the while continuing to monitor other traffic. It's more work, but, in this case, more work makes for better pilot technique.

The approach descent can be accomplished in any one of many different configurations. For simplicity, consider the technique used for a Cessna 152. With only minor modifications, this method can be used for most general aviation aircraft.

Starting at a given heading, altitude, and airspeed, slowly reduce the power to about 1,500 RPM, and hold enough backpressure on the stick to maintain level flight attitude. As the aircraft slows, begin trimming to relieve the pressure on the stick as you put down whatever flap setting you desire to practice for this particular configuration. (My students use everything from clean to full-flap configuration in order to practice for a greater variety of conditions.)

As the airspeed slows to approach speed, keep the power constant and trim for hands-off flight. The nose will start to lower and you will be in an approach descent. Rate of descent is usually controlled with power and airspeed with pitch. In other words, if you see you are descending too fast, you can bring the pitch up a little, add power, or both. You will have to work a little to find the right combinations, especially in rough air. It is worthwhile work, however, because you are really getting to feel the aircraft as well as utilizing both instrument and visual references.

Practice both types of descents straight ahead and in turns in both directions. It will sharpen your skills in airspeed and attitude control and help your descending traffic-pattern turns.

Turns

It has been my experience that turns are high on the list of things many pilots do not know how to do very well. Needless to say, this lack of knowledge sets up problems in many areas of flight that depend on proper turn technique. Turns also happen to be the fourth of the four fundamentals of flight. All four must be mastered before you can even dream of becoming a really proficient pilot.

Unless you happen to be an aeronautical engineer, you probably don't care why an airplane turns. But you had better care *how* it turns because you are the one who is going to have to get the job done. The turning of an aircraft is unique. When you are flying, fly. Don't drive, steer, point, or aim—fly!

Coordination is the key that unlocks the door to a well-executed turn. It takes coordination of the hands, feet, eyes, and mind to properly turn an aircraft, but it is a learned coordination developed over many hours of practice. I have never seen a single student get into an aircraft for the first time and execute a properly coordinated turn. It must be learned.

Although somewhat different from the way you turn your car, turning an aircraft is really not very difficult. The main thing to remember is that you are in an aircraft, not a car. Almost all students have attempted to twist the yoke off of the panel while attempting to turn an aircraft as if it were their car. This happens most often during the early stages of dual instruction and on the ground during taxi when the student attempts to turn. These attempts frequently result in using the prop as a weed mower. Just wander around your local airport sometime during the summer. The aircraft used for primary trainers will be the ones with all the grass dangling from the wheelpants and the greenish tinge on the edges of the prop.

On the ground, an aircraft is taxied at a slow *walk* and is turned through the use of the rudder pedals. The rudder pedals are connected to the nosewheel and provide steering through all ground movements. Brakes should rarely be used for directional control; brakes are for stopping.

In the air, however, an aircraft is turned through the coordinated use of the ailerons, rudder, and elevator. The word *coordinated* is of utmost importance because, if you so desired, you could turn the aircraft with only the ailerons or the rudder. However, this would not produce maximum performance from your aircraft.

An aircraft turns because of a change in the direction of lift—a change from completely vertical lift to lift at a horizontal angle relative to the horizon. This change is implemented with the ailerons producing the bank that then lifts the aircraft in another direction and results in a turn.

A properly coordinated level turn is accomplished when the aileron and rudder are added simultaneously to cause the aircraft to bank to the desired degree. A small amount of backpressure is added to the elevator to keep the nose up and maintain altitude. When the desired amount of bank is reached, everything except the elevator is returned to neutral. The reason the aileron and rudder are returned to neutral is to stop the aircraft in the desired bank. If you were to hold the aileron and rudder in, the aircraft would continue to bank, and neither you nor your trainer aircraft is probably ready for descending rolls.

The elevator should be kept slightly aft of neutral to maintain altitude because an aircraft loses *vertical* lift any time it is in a banked configuration. To make up for this loss, a small amount of up elevator must be held through the turn.

I said an aircraft should be turned through the coordinated use of the controls, even though it could be turned with only the ailerons and rudder. So why make it harder and add both controls when one or the other would do the job? Every pilot should try to achieve coordination to make the aircraft go precisely where and when it is intended to go.

If the ailerons produce the banking motion, then what is the rudder good for? Why is it on the aircraft at all? It is there for two very good reasons. One is to allow the pilot to have control over the yaw, or side-to-side movement, of the aircraft. The other is to overcome adverse yaw produced by the ailerons during the turn entry.

Try this little experiment the next time you fly and you should gain some insight into turns. Fly straight and level and put your feet on the floor. Pick a prominent object or reference point over the nose, such as the smokestack in FIG. 2-8, and roll into a turn without the use of the rudder. The nose will actually seem to go in the opposite direction for an instant, and then it will resume its correct flight path. This phenomenon is called *adverse yaw* and is caused by the drag of the down aileron before it generates enough lift to overcome the resultant drag.

Figure 2-8 shows the result of applying aileron without benefit of rudder in a turn. The turn coordinator ball is moved to the direction of the turn, indicating a *slip*. This can cause a loss of altitude during the turn.

Now, level the wings and start another turn using the same reference point. This time purposely lead the turn with the rudder and then apply aileron. The result will be the plane yawing in the direction of the turn and is known as a *skid*, as shown in FIG. 2-9. With too much rudder, the plane moves laterally to the outside of the turn, as shown by the ball on the right side of the indicator. Hopefully, the result of all this uncoordination will be to instill the habit of monitoring your coordination through the use of both instrument and visual references. In a properly coordinated level turn using outside visual references, the nose of the aircraft will roll about a point and then continue in the direction of the turn (see FIG. 2-10).

Now, combine all the basic maneuvers—straight and level, climbs, glides, and turns—and practice them. Practice climbing and descending turns (left and right, and in different configurations, airspeeds, and flap settings). Get the feel of coordination in your particular aircraft. Only when you master the basics can you continue on and build toward becoming a complete pilot.

SLOW FLIGHT

Slow flight is one of the two maneuvers that separates the real pilots from the aircraft drivers. This maneuver invariably shows whether or not a pilot truly has a feel for his

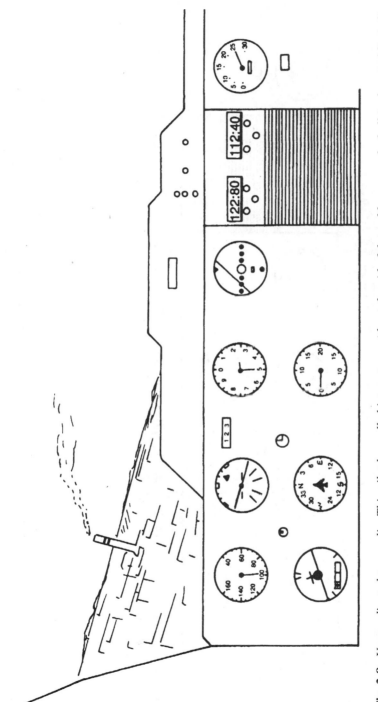

Fig. 2-8. *Uncoordinated turn—slip. This pilot has rolled into a turn without the aid of the rudder—note the ball in the turn/slip indicator is displaced in the direction of the turn, indicating a slipping turn.*

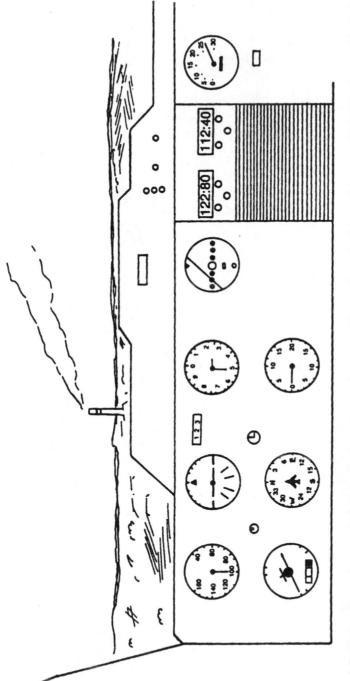

Fig. 2-9. *Uncoordinated turn—skid. This pilot has attempted to turn without aileron (rudder only)—note the ball in the turn/slip indicator is displaced in the opposite direction of the turn, indicating a skidding turn.*

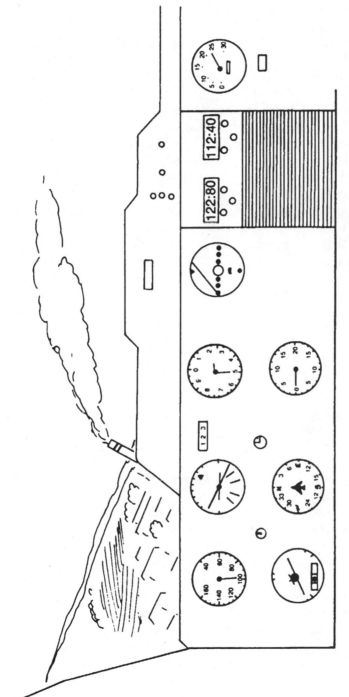

Fig. 2-10. *Coordinated turn. This pilot has rolled into a turn using the aileron and rudder together. Note the ball in the turn/slip indicator is centered, indicating a coordinated turn.*

particular aircraft. Slow flight, if properly performed, is one of the supreme blends of man and machine performing at the very limits of sustained flight. True slow flight, performed at minimum controllable airspeed, is a delicate balance between flying the aircraft with precision and skill or wallowing around the sky hunting your original airspeed, heading, and altitude. It is the fine line that separates controlled flight from a stall.

Stop for a moment and think about the amount of time a pilot spends in slow flight on each and every flight. All of the time an aircraft is not in normal cruise it is in slow flight. It's in slow flight during the climb. It's in slow flight during the descent. And it's most certainly in slow flight all of the time it maneuvers in the traffic pattern during approach and landing. So if a pilot really spends this much time in this flight mode, don't you feel it warrants some real attention? I do.

In slow flight, which I define as anything less than cruise, as the airspeed decreases, control effectiveness also decreases. As the aircraft slows, more control travel is required to obtain the same results that were obtained at normal cruise speed. The reason for this is that there is less airflow over the airfoils as the airspeed decreases. Makes sense. Think of the takeoff roll as the aircraft is gaining speed. When the takeoff roll is initiated, the airfoils don't respond well to control input because the forward speed isn't sufficient to cause much air to flow over the airfoils. However, as the speed is increased, the controls become more effective because more air flows over their surfaces.

Hence, slow flight is more than a mere exercise to practice for a given flight check. It is a safety measure that is used nearly all of the time the aircraft is at less than cruise speed, including the takeoff roll, climbout, descent, and approach to landing. It should be taken very seriously.

Slow flight is no more difficult to master than any other maneuver. Start out at a safe altitude of at least 1,500 feet above the surface and pick a specific heading and airspeed. You want to maintain altitude and heading as you slow the aircraft to begin the practice of slow flight.

As power is reduced to slow the aircraft, backpressure must be added on the elevator to maintain altitude and level flight attitude. If you don't do this, the aircraft will nose over into a descent. As the transition to slow flight is taking place, you will need to utilize the elevator trim to alleviate any control pressure on the yoke. This will also help in nailing down your airspeed. Power, pitch, and trim have to be adjusted to maintain airspeed and altitude. Use whatever power and pitch changes you need to get the job done. One of the foremost problems of good slow flight technique is a sort of shyness many pilots have in the utilization of these controls. They tend to "get behind" the aircraft, and then it's a roller-coaster ride trying to catch up. Stay on top of the situation.

During slow flight, power controls the altitude, and pitch controls the airspeed. Naturally, when you change one, you have to change the other, but only slightly. If you are really keeping up, a small change in power will probably be all you need to counteract a slight climb or descent, just as a slight pitch change can usually cure a small change in airspeed. Remember the trim and let it work for you. If you strive for hands-off flight in all attitudes, your airspeed control will probably take care of itself.

It is also important to take into consideration the effect of torque and P-factor during slow flight. Any time you are in a nose-up, power-on attitude, you have to correct for these two forces. And if you wonder why all of your intended slow-flight headings seem to be somewhere over to the right of where you actually are, remember to correct for torque and P-factor. All it takes is a little right rudder, and the problem evaporates. Use just enough rudder to maintain heading.

Full-Flap Slow Flight

Now that you have the basic slow flight procedure down pat, it's time to complicate matters a little. Try some slow flight using various flap settings. I teach my students to do slow flight with no flaps, full flaps, and everything in between. With this procedure, they develop a good feel for the aircraft in many different configurations. If they're flying an aircraft with retractable gear, so much the better. I teach them to try it both with the gear up and down.

When flaps are added, the nose has a tendency to rise because flaps provide more lift. It will do the same in slow flight, so be ready for it. I have a strong opinion concerning full-flap slow flight: If you think full-flap slow flight is done correctly by slowing the aircraft into the white arc (flap operating range on the airspeed indicator) and then applying full flaps while the aircraft shudders and creaks from the sudden drag of the flaps to slow to the desired airspeed, you are cheating yourself out of all the intermediate configurations that might be very important to you someday. Someone who has never seen an airplane before could learn to perform this method of slow flight in about 15 minutes, but they would be doing it by rote and very little learning would have taken place. Would you want to share your airspace with them?

I maintain that only by practicing slow flight in every conceivable configuration are you really able to get the feel for your aircraft and learn its good points and peculiarities. It will allow you to be comfortable during an approach to landing, while climbing out after takeoff, or any other time you might be maneuvering at less-than-normal airspeed. (FIGS. 2-11 and 2-12.)

Slow Flight Turns

Practicing slow-flight maneuvers, including turns, climbs, and glides, adds to the knowledge of your capabilities and to those of your aircraft. One of the primary objectives during slow flight maneuvering practice is the maintenance of a given airspeed while turning, climbing, or descending. Very little insight has been gained if you reduce the power to practice a descent and fail to decrease your pitch enough to maintain your airspeed. Allowing the pitch to decrease without reducing power only allows the aircraft to go into a dive, not a true slow flight descent. It takes lots of work to truly master slow flight in all attitudes, airspeeds, and configurations; but when you do, you will have taken another giant step toward safe flight technique.

The following example is another sad case from the files of the NTSB. On December 15, 1976, near Patriot, Indiana, a Cessna 150 on powerline patrol became too slow,

Fig. 2-11. *Cessna's Stationair 8 is a flying station wagon!*

Fig. 2-12. *Cessna put a constant speed prop and retractable gear into its 172RG and made a very good airplane into a great airplane with 144-knot cruise capability.*

stalled, and spun in, killing the pilot. The probable causes are listed as diverted attention from operation of aircraft, failure to obtain/maintain flying speed, and stall/spin. The pilot held a commercial license and had 547 total hours with 352 in type.

There are all too many accident reports of the stall/spin variety. I'm certain that most of these accidents could have been avoided with proper training in slow flight and lessons on how to fly and feel the aircraft if your attention is diverted outside the cockpit. Too many people have died because they couldn't fly an aircraft slowly and still maintain a safe margin of control. Intensive practice in slow flight would eliminate a high percentage of these accidents and also make better all-around pilots.

Remember that the art of maneuvering an aircraft with diminished control response is of tremendous importance to all pilots during the takeoff run, climbout, approach to landing, landing roll, and any other time that calls for low-altitude, slow-airspeed maneuvering. Your reactions must be correct and instinctive and honed to perfection. Above all, they must be practiced at regular intervals if you are going to be safe during slow flight.

STALLS

Closely coupled with slow flight techniques are your stall recognition and recovery techniques. As I just said, stall/spin accidents are very high on the list of causal factors for aircraft accidents in IFR as well as VFR flight.

Simply stated, an aircraft cannot stall unless it exceeds its critical angle of attack. Remember that *angle of attack* is the angle between the chord line of the wing and the relative wind. Most airfoils become critical and stall somewhere between 15 and 20 degrees of pitch. Don't exceed your critical angle of attack and you won't find yourself in a stall. It's that simple, and yet people continue to stall, spin, and crash.

Some believe aviation's stall problems stem from a tendency in pilots to believe that stalls happen to other people. Others believe that pilots become complacent when they have not had a close call for a long time. I am inclined to believe that the stall problem stems from a general lack of familiarity with stall recognition and recovery technique. It has been my observation that many pilots fear stalls so intensely that they never learn how to handle them. I see pilots all the time who swell up like balloons during the stall portion of a checkride. Their arms stiffen and their breathing often stops. If that's not fear, I don't know what is.

Another sign of stall fear you might look for in yourself, or others, is the "almost" stall. Many times, when I ask an applicant on a checkride for a certain stall, "all the way through the break," I get an approach to a stall, or I get a prolonged ride in a nose-high attitude with the stall warning horn on. But I don't get a real stall with a full break, the kind you have to take some real corrective action to recover from as the nose breaks over and the wings bank wildly. No, I don't get this kind of full stall from the applicants who fear stalls. I get "almost" stalls.

These folks aren't being careful; they're scared. And if I let them pass with this attitude, I'll bet my last airplane ride that they will never, ever practice a stall again...unless they are forced to for a checkride. The irony of this is that it is the person who is fearful of stalls and avoids them like the plague who most needs the practice in stall recognition and recovery.

To avoid stalling an aircraft because of fear leads to a potentially deadly rationalization. A pilot who fears stalls won't practice them and then, when the recognition or recovery skills are most needed, they aren't there.

If you recognize an approaching stall and take prompt corrective action, you won't have to worry about stall recovery (FIG. 2-13). If you do allow a stall to proceed through the full break, you will have to take quick, correct action using all available procedures. You would be wise to learn and practice both recognition and recovery techniques for all types of stalls.

The Accelerated Stall

The *accelerated stall* occurs when the critical angle of attack is exceeded by abrupt control movement, causing the aircraft to stall at a higher-than-normal indicated airspeed.

Fig. 2-13. *Beware of stalls!*

Just because your aircraft has a V_{so} of 40 doesn't mean that there is no way it can be stalled at a higher speed.

You don't necessarily have to be nose high to stall it, either. If you exceed the critical angle of attack, you can stall your aircraft going straight up, straight down, or straight and level. Remember, in an accelerated stall, the stall occurs at a higher-than-normal airspeed and usually requires some very rapid control movement to induce the stall. But it *will* stall.

Everything has its limits and so does your aircraft. What is the highest airspeed at which you can stall your aircraft *safely?* The answer is the *maneuvering speed* (V_a)—the speed at which you can apply full, abrupt control travel without causing structural damage to your aircraft. Anytime you are at or below maneuvering speed, you can stall your aircraft without worrying about whether you or your wings are going to get to the ground first. It is a built-in safety factor tested by the aircraft manufacturer and certified in your aircraft's operations limitations. But don't rush out and try it. Take my word for it for now; get a little more acquainted with stalls, and then decide if you really want to give it a go. Personally, I never have, except in aerobatic aircraft.

To practice accelerated stalls, first climb to minimum stall recovery altitude of 1,500 feet above the ground, and then climb another couple of thousand feet for mistakes. Clear the area by doing a 90-degree turn in each direction (or a 180, whichever makes you happy). Look for other aircraft while doing the turns. Too many people go through the motions of clearing turns and use the time to unwrap a piece of gum or adjust the ventilation in the cabin.

The entry speed for the accelerated stall should be no more than 1.25 times the unaccelerated stall speed in a clean configuration: V_{sl} times 1.25. If your aircraft has a V_{sl} of 60 knots, this would work out to an entry speed of 75 knots ($60 \times 1.25 = 75$). This rather low speed provides a margin for safety because the load factor at the time of the stall will be lower if it begins at 1.25 V_{sl} rather than up near maneuvering speed. The stall is less violent and causes less stress on you and your aircraft.

After you arrive at altitude and clear the area, slow your aircraft down to entry speed, and while maintaining constant altitude and a low power setting, roll into a coordinated 45-degree bank. Rapidly increase the pitch as you hold your altitude and the aircraft should stall at a higher-than-normal airspeed. If properly executed, it will stall quickly and break rather sharply. If you turn more than 90 degrees during the execution of the accelerated stall, you aren't pulling back fast enough on the yoke and probably aren't getting a true accelerated stall. You don't have time to eat a hamburger while doing an accelerated stall properly. It happens in a hurry.

Recovery from the accelerated stall is as with all stalls:

- Reduce the pitch to break the stall.
- Add power, if available.
- Level the wings and return to normal flight attitude.

Complete these procedures in that order and quickly. The stall recovery can be initiated at any of the normal recognition cues peculiar to stalls (or impending stalls). These cues are common to all and include the warning horn and/or light, decreasing control effectiveness, buffeting, and finally, the break. The most sinister thing about the accelerated stall is that these cues happen much more rapidly than is the case with most stalls. But if you practice recovery at all the different cues, you will be ready should you ever get into a situation that calls for prompt action.

Leveling the Wings

While on the subject of stalls and stall recovery technique, I would like to bring home a point that I have observed over many years of flight instruction as one of the most recurring problem areas in stall recovery-leveling the wings. It's number three on the list (last), but it's there for a very good reason. When an aircraft stalls, the airfoils are what actually stall. The wings and horizontal stabilizer stall and remain stalled until you release some back pressure to once again allow air to flow over these airfoils in a normal pattern. No matter how hard you try, you will not be able to level the wings with the ailerons until the stall is broken.

But, it's a natural tendency to try, by using the ailerons, to pick up a wing that has suddenly dropped off. I have seen some students literally twist the yoke to the point I feared they would damage the controls trying to pick up a down wing that was still stalled. It won't work for the aforementioned reasons. What, then, will work? *Rudder*. The rudder remains effective long after the other control surfaces have become useless and should be used to pick up a down wing during stall recovery. Use the rudder pedal opposite the down wing to initiate wings-level attitude during the initial stall recovery.

I'm not recommending the usage of the rudder to the exclusion of your other flight controls during stall recovery. I simply mean it should be implemented first. Then, normal recovery is accomplished through the coordinated use of all available controls and power.

Takeoff and Departure Stall

The *takeoff and departure stall* usually occurs due to lack of pilot attention to pitch control during this critical portion of flight. It is often caused by an overzealous pilot pitching the aircraft up to such an extent that the stall is unavoidable. At any rate, far too many accident reports carry the final notation: "Failed to obtain/maintain flying speed."

The takeoff and departure stall can occur any time the pilot pulls the aircraft up too steeply or perhaps becomes engrossed in something other than flying the airplane—like showing off. If you want a blueprint for a prime example of an accident just waiting to happen, read on.

Several years ago, at a nearby airport a wealthy businessman was taking flying lessons from the local FBO. As the man approached the time he knew he would solo, he made a pact with his wife that on the day he soloed he would fly over their house and she was to come outside and videotape his flyovers. Of course, his instructor knew nothing of his plans.

The day of his solo came and instead of remaining in the traffic pattern as he had been told to do, the businessman turned away from the airport and headed over town. He was gleefully buzzing his house and his wife had the videotape rolling when it happened. After buzzing his house at a very low level, he pulled the airplane up at too steep an attitude, stalled, rolled over and crashed nearly straight down in his own neighborhood. Of course, it was fatal. Yes, inattention at so critical a time with such an inexperienced pilot can be the last thing a person may ever do. Literally.

In order to practice the recognition of and recovery from this stall, take your aircraft to at least 1,500 feet above ground level and again add a couple of thousand feet for error. Clear the area, as always, before practicing stalls. This particular stall is initiated at liftoff speed, so retard the throttle well below cruise RPM and maintain a constant altitude as the aircraft slows to liftoff speed. As you reach liftoff speed, advance the throttle to full takeoff or climb power and increase the pitch simultaneously, not allowing any acceleration of the aircraft. After the desired pitch is established, all controls are returned to neutral except for the rudder, which must be used to overcome torque and P-factor-the right rudder, that is. Recovery can be initiated at the first indication of the impending stall or after the full break.

The takeoff and departure stall should be practiced from a straight climb as well as climbing turns in both directions. When you practice it while turning, use a moderate bank of 15 to 20 degrees. Notice how the cues in this stall seem to come one at a time. You should be able to recover at any one of the cues. In fact, to the competent pilot, these cues almost scream for action on the part of the pilot. Notice them, feel them, and practice them. Recovery will be normal except for the power. Because you already have a high-power setting, reduce the pitch to break the stall and level the wings using coordinated control forces. Then, fly the aircraft out of the stall with as little altitude loss as possible, and set up a normal climb.

Another example can be taken from NTSB File # 3-3612. On December 12, 1976, near Malta, Montana, a Piper PA-18 flown by a private pilot with 1290 total hours stalled in a low turn and crashed. Fortunately, it was nonfatal.

The NTSB lists the probable causes: diverted attention from operation of aircraft, failure to obtain/maintain flying speed, lack of familiarity with aircraft, first flight with skis, stalled in turn.

Approach-to-Landing Stall

The *approach-to-landing stall* often occurs on the turn from base leg to final approach, but it has been known to happen when a pilot attempts to stretch his glide on final without sufficient power to maintain altitude. The turn from base leg to final approach is the prime area for this stall because pilots often start their turn to final too late and are tempted to add a little inside rudder and some back pressure to tighten the turn. Seeing that this helps a little but that they are still going to overshoot the runway centerline, they sneak in a little more rudder and backpressure to tighten the turn even more. Often they use opposite aileron to try to keep the turn from overbanking due to the effect of the rudder and backpressure. They are so intent on making the runway on the first try, rather than going around and flying a better pattern, that they fail to notice the cues the aircraft is giving them. It is the perfect setup for an approach-to-landing stall.

Approach-to-landing stalls are executed from normal approach speed and should be practiced from straight glides as well as gliding turns (the latter simulates the turn from base to final). Clean and full-flap configurations should be practiced because you might be called on to make various approaches in various configurations during the conduct of actual flight. Also, these stalls should be practiced at an altitude higher than the minimum recovery altitude of 1,500 feet above the ground because you will be in a descent during the maneuver.

After carefully clearing the area, smoothly begin to retard the throttle to the approach power setting. Maintain your altitude as you slow to the approach speed. During this time, the flaps should be lowered to landing position (assuming the aircraft is so equipped). When normal approach speed is reached, initiate a descent. After descending about 200 feet, begin to smoothly increase the pitch as the power is reduced to near idle. Continue to increase the angle of attack until the stall occurs. The stall usually occurs with very little nose-high attitude. Pulling the nose up too sharply ruins the intent of the maneuver. Because this stall almost always happens slowly and gradually, you should be able to recognize and feel the cues as they come one at a time.

Recovery from an approach-to-landing stall can be initiated at any point, up to and including a full break. Recovery is much like a full-flap go-around from a landing approach. Relax backpressure and smoothly add power as you level the wings and transition to best-angle-of-climb or best-rate-of-climb airspeed. While you are doing this, the flaps should be brought up to the manufacturer's recommended setting for a go-around. In most aircraft, the flaps are *not* brought up all at once because that can cause a momentary sink, which could prove problematic at a very low altitude.

As I stated before, this stall should also be practiced from gliding turns in both directions, using a moderate bank of 15 to 20 degrees. All other procedures remain the same.

STEEP TURNS

As one of the most difficult maneuvers required for a private pilot certificate, let's consider steep turns as a maneuver to grow into rather than one in which the student pilot should attack. Attempts to conquer steep turns early in one's flight training often results in a rather poor performance of the maneuver. The steep turn is another maneuver that demonstrates whether or not the pilot is truly the master of his aircraft (FIG. 2-14). A steep turn requires an advanced sense of coordination and timing. Maintaining altitude and orientation are only two of the problems this maneuver presents. Steep turns take the aircraft to the opposite end of the performance spectrum from slow flight. However, the possibility of an accelerated stall still exists as the aircraft reaches a 50- to 60-degree bank and large doses of backpressure must be applied to help maintain altitude.

The aviation industry seems to be divided on the subject of what really controls altitude during a steep turn. Many reputable pilots contend that backpressure alone controls the altitude. I tend to agree with them to a point. If, as you roll into a steep turn, you feed in *exactly* the correct amount of backpressure, I agree it does control the altitude. However, pilots who can do this consistently are few and far between. During the roll-in, most pilots usually add too little or too much back pressure versus bank and wind up chasing

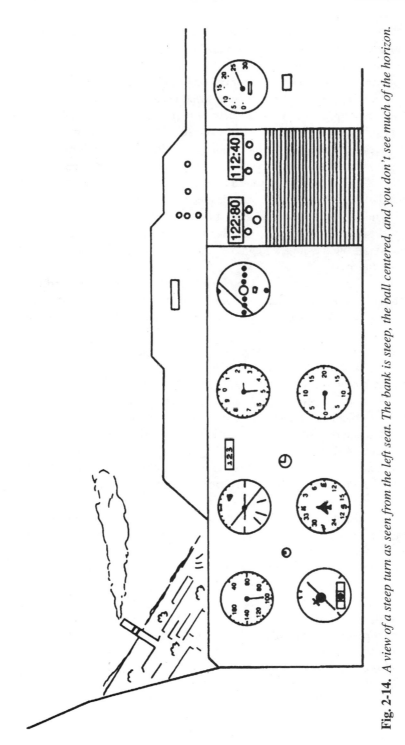

Fig. 2-14. *A view of a steep turn as seen from the left seat. The bank is steep, the ball centered, and you don't see much of the horizon.*

the pitch all the way around the turn, resulting in what I call 720-degree vertical S-turns. It does little for training and even less for your stomach. The method I utilize incorporates pitch and bank control to maintain altitude.

After clearing the area, set up straight-and-level flight below maneuvering speed. (Steep turns should be done at or below maneuvering speed because of the possibility of momentarily high G-loads caused by the pitch changes required to maintain altitude.) Pick a heading and altitude at which to begin the steep turn. After completing a 360-degree or 720-degree steep turn, this will be the heading and altitude at which you want to wind up. Then, begin the roll into the steep turn at a higher-than-normal rate. As the bank reaches about half the desired maximum, 25 to 30 degrees, smooth application of backpressure is initiated as the roll continues to the desired amount of bank. If your backpressure input versus bank angle is relatively close to the desired amount, altitude can then be controlled by slight changes of bank to either increase or decrease the vertical lift. At the same time, small pitch changes might be necessary to complement the bank and aid in proper coordination.

Unless you're a master at steep turns or enjoy high G-loads and feeling centrifugal force to the point that your fillings are pulled from your upper teeth, the vertical lift component can be taken care of through small changes in bank instead of large changes in pitch. Try it.

Because a steep turn usually produces about two Gs, the airspeed will decay due to the higher load factor. Power should be added as needed to help keep the airspeed up. Remember, as the bank increases, so does your stall speed. Therefore, power application during the steep turn should help decrease the chances of an accelerated stall.

The most common problem associated with steep turns is uncoordinated entry and recovery technique. Pilots entering with too little backpressure for a given bank lose effective lift due to the steep bank and enter a descending spiral. Pilots entering with too much back pressure for a steep bank find themselves in a tight, climbing turn, Proper aileron and rudder input and the correct amount of back pressure help assure a good entry into a steep turn.

To keep a steep turn coordinated, a slight amount of opposite aileron must be used to overcome the over-banking tendency as the bank reaches the 50- to 60-degree range. *Do not* use top rudder to correct the over-banking tendency.

Recovery from a steep turn is as with all turn recoveries, only more control pressure is required to make it come out in a coordinated fashion. Since you turn at a high rate, it takes more rudder application than normal to overcome the turn and also the adverse yaw effect of the ailerons as they are applied to reverse the bank. Also, as you proceed through the rollout, relax the backpressure to prevent a climb after the turn is completed. In other words, as you roll back to level flight, if you don't release the back pressure you held through the maneuver, you will increase your altitude at the end and ruin a maneuver which might have been otherwise alright.

3
Ground Track Procedures

IN ALL PHASES OF AVIATION, YOU ARE AT THE MERCY OF THE WIND from the time you lift off until the time you touch back down. Take it even one step further—you're contending with the wind from the time you untie your aircraft until you tie it safely down again. For this reason, it is imperative that you master ground track procedures very early in your flight experience. It makes takeoff and departure, approach and landing, getting from point A to point B, and everything else relevant to flight much easier to accomplish if you understand and can execute ground track procedures properly. To me, it is flight. It is important to go where you want to, when you want to, and to know the wind will be nothing more than a minor inconvenience and not something to fear. You must learn to respect wind, deal with it, but not fear it.

TURNS ABOUT A POINT

The *turn about a point* is probably the most familiar ground track maneuver to most pilots. Almost everyone has had to, or will have to, demonstrate this maneuver on one checkride or another. All too many pilots work very hard to get ready for a checkride and then after completing that ride, forget everything they learned about a maneuver or group of maneuvers. This is often the case with turns about a point.

Turns about a point, as well as all ground track maneuvers, are entered downwind. The primary reason for the downwind entry is safety. Because the angle of bank is proportional to the groundspeed, the faster the groundspeed, the steeper the bank. The slower the groundspeed, the shallower the bank. Therefore, if the turn about a point is entered downwind, the initial bank will be the steepest encountered in the maneuver. It prevents a pilot from making the mistake of entering the maneuver upwind (into the wind) at a steep bank, only to find that to hold the desired radius, the bank must be increased to a point that might exceed the capabilities of the pilot or the aircraft.

All ground track maneuvers incorporate an infinite number of bank angles, but the three most used reference banks are shallow, medium, and steep. Once you have mastered the basic rules of bank as they apply to ground track, you should have no difficulty with any ground track maneuver. The basic rules of bank are:

- Steep bank downwind (wind behind you).
- Shallow bank upwind (wind in front of you).
- Medium bank crosswind (wind from your side).

Remember that the wingtips point at the reference point only when you are directly upwind or downwind. Everywhere else your wingtips will be either ahead of or behind the reference point. Don't try to keep the wingtips on the point or you will make a very egg-shaped circle. Crabbing is important at all times, not just directly upwind or downwind.

The turn about a point can be a very large or very small circle. It all depends on how close you are to the reference point. The closer the aircraft is to the reference point, the steeper the banks will be, and the smaller the circle. The farther away from the reference point the aircraft is, the shallower the bank, and the larger the circle. And remember, banks are relative. Certainly a 45-degree bank is steeper than one of 15 degrees, but it is only three times as steep, whereas a bank of five degrees is five times as steep as a bank of one degree. So think of your angle of bank as one relates to the other, not in terms of 45, 15, 10 degrees, etc. Think of them as shallow, medium, and steep.

To properly execute a turn about a point (FIG. 3-1), you first need to attain the proper altitude. The FAA recommends you be from 600 feet to 1,000 feet above any obstacles as you choose a point about which to pivot. Make sure the point is stationary. I once had a student try to do a turn about a point around a car. And the car was going about 60 MPH down the road—truly an advanced maneuver for which the student was not ready.

As you enter downwind and cross abeam of the point, the bank will be initiated. It should be a coordinated turn resulting in a steep bank. The steep bank will be made progressively shallower through the first half of the circle. As the crosswind point is reached, the bank will be medium, and it will also be the point of maximum crab angle since you will be directly crosswind. As the aircraft proceeds from the crosswind to the upwind position, the bank will get progressively shallower until you arrive at the upwind position with the shallowest bank of the entire maneuver. You now have turned 180 degrees. At this point, the entire banking procedure reverses and gets progressively steeper as you go

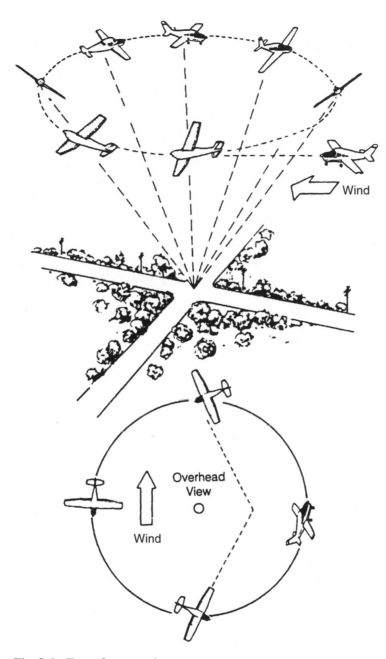

Fig. 3-1. *Turn-about-a-point.*

from upwind to crosswind. Proceeding from crosswind, the bank continues to steepen until you arrive at the downwind starting point and again have the steepest bank of the maneuver.

If there is no wind, theoretically, the bank will be constant throughout the 360-degree turn. Therefore, it is probably best to practice this maneuver in a steady breeze. A gusty wind condition can cause you some difficulty in acquiring a feel for changes in ground speed and resultant banks required to form a constant radius about the point. Remember to strive to hold a constant altitude throughout the maneuver.

And do not become so engrossed in your maneuver that you forget to look out for other aircraft. I remember flying at 6,000 feet giving some aerobatic training and looking down to see what I at first thought was two students dog-fighting. They were far below us, in unison, going around a circle in perfect symmetry. As I watched them for a minute, it began to dawn on me that they were, in fact, both doing turns-around-a-point around the same point! And they never saw each other! Imagine the potential for tragedy. I quickly called one of the aircraft on the radio and instructed them to break to the right since they were in effect flying in formation without the benefit of seeing the other aircraft. Live longer, look around.

S-TURNS

The *S-turn* is a fine training maneuver, requiring the same bank techniques incorporated in the turn about a point. The S-turn differs from the turn about a point in that instead of flying a complete 360-degree circle, you make a series of 180-degree turns along a road or other straight landmark. Once again, crab angle, along with shallow, medium, and steep banks, are your primary guides in obtaining the desired track over the ground.

Begin at about 600 feet above obstructions and over a road that lies perpendicular to the wind (FIG. 3-2). Enter downwind for the same reasons mentioned for the turn about a point. As you approach the road, decide how large you intend the half circles to be. Remember, the points of maximum distance from the road should be equidistant on both sides of the road. This equidistance helps maintain the symmetry of the maneuver.

The S-turn maneuver begins when the aircraft is directly over the road. At this point the bank is initiated, and is your steepest bank because the groundspeed is the fastest here. As the aircraft continues to the crosswind point, the bank gradually shallows to become medium and you encounter the maximum crab angle. Turning to upwind, the bank gradually decreases to the most shallow as the aircraft reaches the 180-degree point. It occurs just as the aircraft reaches the road, with the aircraft exactly perpendicular to the road.

The bank is reversed for the turn in the opposite direction. The aircraft is still upwind, so the shallow bank should be held until the aircraft turns far enough to cause you to gradually steepen the bank when you arrive at the crosswind point, with medium bank and maximum crab angle. From crosswind, continue to steepen the bank gradually until you arrive back over the road with the steepest bank for this half of the maneuver. You should arrive back over the road just as you complete the 180-degree turn, not before or after. The two half circles just completed form one complete S-turn.

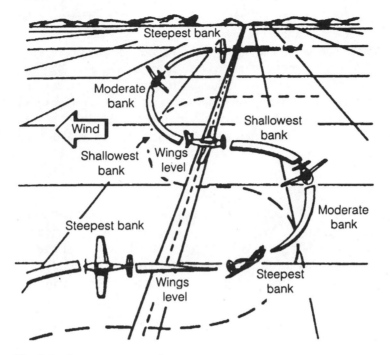

Fig. 3-2. *S-turn over a road.*

Of course, you don't have to stop with just one S-turn. In fact, it's probably better practice to put together several in one direction and then reverse your course and go back down the road in the opposite direction. Be cautious and don't become so engrossed that you forget to watch for other aircraft. Seldom are you as alone as you think you are. In fact, given the scope and area of the sky, it is amazing how many midair collisions and near misses we have in the U.S. each year.

The S-turn may be made any size the pilot desires. The only thing governing the size of the half circles is your initial bank. The steeper the initial bank, the smaller the circle. The shallower the initial bank, the larger the circle. The symmetry of the S-turn is kept uniform by going twice as far *down* the road as you go *out* from the road. This symmetry ensures a perfect half circle.

The list of common errors in the execution of an S-turn would be headed by the pilot hurrying the turn from upwind. At this point, the groundspeed is the slowest and there is a tendency not to hold the shallow bank long enough for the aircraft to fly away from the road as far as it was on the downwind side. The best way to overcome this is to pick reference points of equal distance from both sides of the road and then, without cutting corners, fly to them.

Crossing the road before or after the maneuver is completed is another common error. The aircraft should cross the road just as the 180-degree turn is completed and roll into a turn in the opposite direction. Since most of the time you will be flying with your head out of the cockpit for visual reference, a brief glance at the altimeter every so often confirms any altitude change you might not be aware of.

The S-turn is one of the better coordination exercises because you have to use visual as well as instrument reference, feel for the aircraft, and all your senses, with the possible exception of taste and smell. While you are doing all this turning from one direction to the other, always remember to watch for the other aircraft as well as monitor headings, altitude, airspeed, and bank. This constant monitoring helps improve the ability to think and act quickly and accurately. When taken seriously, the S-turn is a very exacting and demanding maneuver worth perfecting.

RECTANGULAR PATTERNS

The rectangular pattern is another ground track maneuver that involves not only varying the bank to correct for wind drift, but a great deal of crabbing into the wind to keep the aircraft on the desired path (FIG. 3-3). In the S-turn and the turn about a point, a little crabbing was done as the aircraft reached the crosswind point. But in those maneuvers, the aircraft was turning throughout the entire maneuver. In this ma-

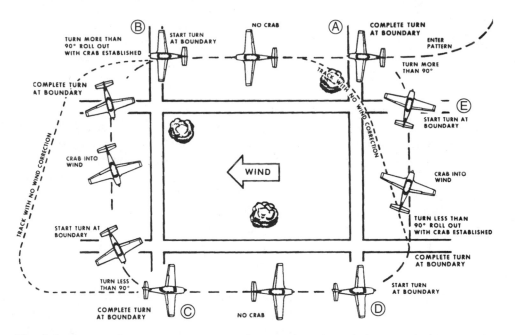

Fig. 3-3. *Rectangular pattern: A) Enter downwind at desired distance from rectangle; B) Downwind to crosswind turn, steep-to-medium bank, more than 90-degree turn, crab as necessary to maintain ground track; C) Crosswind to upwind turn, medium-to-shallow bank, less than 90-degree turn; D) Upwind to crosswind turn, shallow-to-steep bank, less than 90-degree turn, crab as necessary to maintain ground track; and E) Crosswind to downwind turn, medium-to-steep bank, more than 90-degree turn. (Note: All turns begin and end perpendicular to field corners.)*

neuver, there is a lot of straight-and-level flight with crabbing the only means for correcting wind drift.

The rectangle pattern is flown parallel to, and equidistant from, the field or group of fields used to make up the rectangle. As in all ground track maneuvers, the entry is made downwind and the altitude should again be about 600 feet above obstructions.

Fly the aircraft to the point that is parallel to the downwind corner of the field. You are flying parallel to the field on a downwind heading. You will have to crab as necessary to maintain equal distance from the field. At the downwind corner of the field, make your first turn. Because the turn is from downwind to crosswind, the bank should begin steep and gradually shallow out to a medium bank as the aircraft reaches the crosswind point. The turn should be complete with all crab established at a point directly parallel to the corner of the field. This first turn is more than 90 degrees because of the crab angle required to maintain a straight ground track. Continue on, crabbing to hold a constant distance from the field on the downwind side, until you reach a point directly parallel to the next corner of the field.

Upon reaching the second corner, the aircraft is still crosswind. The initial bank is medium, gradually decreasing to shallow as you reach the point where you are into the wind and parallel to the corner of the field. This turn is less than 90 degrees because the turn began with the aircraft in a crabbed situation, pointing slightly into the wind and it ended up pointing directly upwind (into the wind). All of these points assume the wind was on your nose. If not, you will have to crab, as necessary, to hold your constant distance from the field.

At the third corner, the aircraft is pointed into the wind, so the initial bank should be shallow and gradually increased to medium as the aircraft reaches the crosswind position. This turn is less than 90 degrees because the turn was started into the wind. The turn is completed crosswind, requiring a crab angle to hold your flight path parallel to the field.

Upon reaching a point parallel to the fourth corner, remembering you are still crosswind, start out with a medium bank and gradually increase it to steep as you turn downwind. This turn should be more than 90 degrees because you started the turn with a crab into the wind, and the turn ended up directly downwind. If you are directly downwind, no crab should be required to hold your pattern an equal distance from the field on this leg. This process completes one circuit of a rectangular pattern.

As you have probably already surmised, the rectangular pattern has a direct relationship to the normal traffic pattern. Since all good traffic patterns are rectangular, you must learn how this maneuver transfers to the traffic pattern. Many pilots learn how to execute a rectangular pattern and then forget they have ever done one upon entering the airport traffic pattern. They fail to allow for wind effect and fly very erratic traffic patterns. It is a prime example of learning a specific maneuver and then not utilizing it in a practical situation.

I have had students who, after an entire period of instruction on ground track technique, have entered the traffic pattern and were blown so far off course that I wanted to implant a little knowledge into their heads with a sledge hammer. They simply did not apply the basic technique learned during the lesson to the situation at hand. This is one of the most frustrating things for an instructor. You can teach a person how to do a maneuver, but you can't teach him how to think. This type of student usually finds the postflight discussion brief and to the point. The lesson learned out in the practice area is of no value until you can apply the learned skills to a real situation. Only then can you be sure that real learning is taking place. Instructors do not teach all the maneuvers simply to give the examiner something to look at on the checkride. They must be learned, remembered, and put to use throughout your flying career. They must become habits.

If the preceding thoughts on the subject of the importance of proper ground track procedures have not moved you to improve your habits, think about this example from NTSB file # 3-3762. On September 26, 1976, near Steamboat Springs, Colorado, a Piper PA-23 took off on a flight to Denver. They never made it out of the pattern. During takeoff, the aft cargo door came open and the pilot decided to return and land. The pilot, 32 years of age with 4,686 hours of flight time and carrying five passengers, lost control of the aircraft while turning from base leg to final approach. The NTSB lists the probable causes as diverted attention from operation of aircraft; failure to obtain/maintain flying speed. In the remarks section it says that witnesses saw the aircraft turn to final and the bank steepen, followed by a sharp nose drop. Six fatalities. The picture that comes to mind is one of a pilot so intent on something other than flying the aircraft that the wind blew them too close in on the downwind leg. It resulted in the pilot overshooting the turn to final. He probably tried to correct it by tightening the turn but then stalled and ended up in disaster. This picture might not have been the case in this particular instance, but I have seen similar happenings time and time again, though they don't always end up so tragically. Your very first and foremost responsibility is to fly your aircraft; distraction leads to destruction.

4
Takeoffs and Landings

A FRIEND OF MINE GAVE ME A COFFEE MUG ON WHICH WAS INSCRIBED, "Flying is the second greatest thrill in life—landing is the first!" I guess this might be true if you are a bit fearful of flight and are indeed happy to be back on good old terra firma. But to all too many pilots, the takeoff and landing are the sum total of aviation. They try to show their proficiency as pilots by the way they take off and land.

If you have carefully read the pages of this book leading up to this point, you hopefully have received the message I am trying to impart—this attitude is not correct. There are so many things that lead to good overall piloting ability. Taking off and landing happen to be two of the pieces that make up the overall picture. Granted, taking off and landing are important, but to do so with great skill and confidence, you must first master a host of other important and related tasks. When these tasks are mastered, then you are ready for work in the traffic pattern, but not before.

With today's tricycle gear aircraft and miles of hard-surface runways, the normal takeoff is one of the easiest maneuvers to master. Almost all you have to do is point the aircraft straight down the runway, apply power, keep it straight until rotation speed is reached, and rotate. It's that simple. Or is it? If you are a robot, it is. If you are a pilot, there's a little more involved, such as knowing how to straighten it out if things suddenly go astray and remembering to keep one eye on the oil pressure and airspeed indicator while keeping the other eye on runway alignment of the aircraft, all at the same time. It goes on and on. (See FIGS. 4-1 and 4-2).

Fig. 4-1. *A simple airport.*

Fig. 4-2. *A more complex airport.*

GROUND OPERATIONS

Before you can take off, you have to be able to get your aircraft safely started and out to the runway. While this may sound rather simplistic, it can be a source of problems to the careless or uninitiated.

The first and foremost considerations in ground operations are common sense and courtesy. For instance, when you are ready to fire up your engine, you always holler "Clear!" don't you? But do you wait a few seconds after you call out so some poor boob can actually move himself out of the propeller area? Applicants on a checkride are sometimes nervous and are especially susceptible to calling "Clear!" as they are turning the key to start the engine. I'll tell you this: Anyone in the way of the prop with one of these guys will be looking for his parts over a wide area. This is one example where thoughtlessness can turn deadly. Here's another:

An acquaintance of mine was in the process of starting his aircraft on his ranch, far away from the potential problems we mere mortals face at the local airport where we have to share space with the rest of humanity. As he prepared to start up his aircraft, it became clear to me that he wasn't going to clear the area before turning the starter switch. I asked him if he shouldn't clear the area even though we were in the boonies. He told me that, "Nah" he never needed to verbally clear at the ranch. There were only so many people around and he had them all accounted for. This particular day he started his aircraft and threw his favorite dog about 50 feet, straight into doggy Heaven. We'll never know if old Scruffy would have responded to a loud call, but I'll bet he would have come around just to see what his master wanted. And you can believe any person over five years old will move in a hurry if warned of an impending startup. The point is that people and animals certainly deserve the chance to get out of the way, so slow down and give safety a helping hand.

If there's anyone standing behind you, they might also wish to move rather than be showered with a propeller blast, rock chips, dirty water, oil, or who knows what. A small amount of courtesy goes a long way towards a safe flight. And a safe flight literally begins when you strap into the aircraft.

Taxi Procedures

As you begin to taxi, move away with only enough power to initiate forward movement, and then retard the throttle to a point that lets you taxi at about a fast walk. I said a fast *walk*. I have ridden with some pilots who taxi faster than a cheetah can run. And it gets pretty interesting when they come to a corner. Taxi only as fast as you are comfortable with, and never attempt to turn a corner too fast in an aircraft—they can, and will, tip over. Wouldn't you love to try to explain that at the next company party?

As you taxi, it is of vital importance that you position the controls to minimize the effects of wind as it relates to lift, directional control, and yaw tendency. (See FIG. 4-3.) The idea is to position the ailerons and elevator to generate the least amount of lift under the wings and tail during taxi, especially when the wind is from the rear. You want the wind to hold the wings and tail down rather than create any lifting action that could overturn you during this critical phase. Remember that aircraft are not at home until they become airborne. In fact, most aircraft are rather awkward on the ground.

During the run-up prior to takeoff, you turn your aircraft directly into the wind. This will do two things for you: 1) It will provide maximum cooling for your engine; and 2) It will maximize the aerodynamic flow of air over your control surfaces. In fact, if you

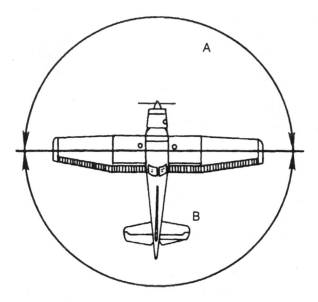

Fig. 4-3. *Selecting control positions during taxi. Imagine a line running through your aircraft from wingtip to wingtip, (the lateral axis): A) If the wind is coming from in front of the imaginary line, turn your control wheel towards the wind direction and hold neutral or up elevator; B) If the wind is coming from behind the imaginary line, turn your control wheel away from the wind direction and hold forward elevator.*

have a bit of a breeze, when you test your controls for freedom of movement, the aircraft will react just as it would if airborne. Pull back on the yoke and the nose should rise slightly, etc.

Lastly, any run-up should be completed with the nosewheel pointed straight. If you stop during a turn, leaving the nosewheel crooked, the thrust from running up the engine can cause the bearings to get out of round. Then, when you take off or land, the nosewheel can shimmy very badly. If you have flown a plane in which the nosewheel vibrates violently at some point, you can usually attribute the cause to the plane having been run up with the nosewheel in a cocked position.

NORMAL TAKEOFF

If you have learned your lessons well out in the practice area and can put to use the principles of slow flight, stalls, straight and level, and all the rest, the normal takeoff will hold few surprises for you.

As I taxi onto the active runway, I always do a little mental jog I call my *FFT check*. It's a last check of the three most often overlooked items on your pretakeoff checklist. I call them the "killers" since a takeoff with one of these items set incorrectly can lead to dire consequences. FFT stands for fuel, flaps, and trim. As I think "fuel, flaps, and trim," I carefully check each one to be sure it is indeed in the takeoff setting. Some pilots have come to a sorry realization at a very inopportune time that one of these wasn't set correctly. One memorable event that cost many lives occurred to the crew of an MD-80 at Detroit, who attempted a takeoff with the flaps in the wrong position. This very preventable human mistake could have been avoided with a simple last-minute check that the fuel was on and that the flaps and the trim were set correctly.

When cleared, taxi onto the runway at the end. Don't waste 200 yards of that precious runway weaving back and forth to line up with the centerline. Go right to it. Line the longitudinal axis (nose-to-tail axis) with the centerline of the runway, and smoothly apply full power. You might need a little right rudder as the power is applied to overcome torque. Feel the controls begin to become effective as your speed increases. Glance quickly at the oil pressure and airspeed indicator, then get your eyes back outside where the action is. If you keep your eyes in the cockpit too long, there might be more action waiting for you out there than you want.

As the aircraft gains speed and the controls become more effective, use small control inputs to maintain your line straight down the runway. One of the most common errors in primary flight is the tendency to overcontrol, especially the rudder. However, be sure to use whatever it takes. Don't go the other way and be too timid with the controls either.

When you reach the rotation speed for your particular aircraft, smoothly apply a little backpressure and lift off (Fig. 4-4). The aircraft should be rotated so that you will be at an angle that produces a climb at about V_y (best rate-of-climb speed). If you find this a

Fig. 4-4. *Everything checked, lined up, and ready to go.*

little difficult at first, don't feel alone. The proper amount of rotation will come with a little time and practice.

Okay, now you're airborne and climbing out at V_y. Trim the aircraft to help maintain airspeed while you glance around to verify that you are climbing straight out from the runway centerline. Any drift should be corrected for by the use of the crab technique. Remember that you learned to correct for wind when you learned ground track procedures. Now is the time to put that knowledge to practical use.

The FAA recommends climbing to at least 500 feet above the ground before turning out of the pattern. Another suggestion is to climb straight ahead until you reach the end of the runway. Unless otherwise directed by the tower, do whichever comes *last*. The reasoning behind this is to try to get better standardization from one airport to another.

NORMAL LANDING

There's an old axiom in aviation that says, "A good landing is usually preceded by a good approach." This statement is still true today. A normal landing, the one you learn before going on to the other advanced types of landings, is like all other landings. It is merely the end result of a transitory period from the approach, through the flare, to ground contact. But before you can execute this transition to a landing, you have to arrive over the runway threshold at approximately the correct altitude and with proper alignment. Therefore, you first need to execute a good approach before you can attempt a good landing. And you can either *land* or you can *arrive*. Arrivals usually bring forth much laughter from your flying comrades and another grey hair for your instructor. Sometimes arrivals strain the landing gear, bend props, flatten tires, etc.

The normal landing begins long before the actual touchdown. It starts with your preplanning for your pattern, airspeed, traffic spacing, flap usage, and all the basic flight techniques you have previously learned. "A good landing is no accident," says the FAA.

On the downwind leg, you should go through the prelanding checklist, set your aircraft up at the proper distance from the runway, check traffic, and note that you are at traffic pattern altitude. You should be exactly *on* your altitude, not close to it. About halfway down the runway on the downwind leg, pull on the carb heat so it has time to work. Remember that the heat from the exhaust warms the carburetor when you pull on the carb heat, and the air is not exactly hot enough to melt steel. Even a blowtorch takes some time to melt ice.

When you are directly abeam of the approach end of the runway, reduce your power to an approach setting and trim your aircraft to maintain the recommended approach speed. After you have set up your glide and the airspeed is definitely in the white arc (flap-operating range), lower the first 10 degrees and retrim the aircraft as necessary to maintain approach speed.

Now comes one of the most important decisions you have to make to keep your traffic pattern uniform: When do you turn onto base leg (FIG. 4-5)? A good rule of thumb is to turn base as soon as you arrive at a point where the end of the runway is at a 45-degree angle behind the wingtip nearest the runway. This procedure prevents you from getting into the bad habit of turning over a certain tree, house, bend in the road, etc. Remember that particular reference point won't be available at another airport.

After you turn to base leg, another important judgment must be made: Are you too high, too low, or just right? This is called the key point in the approach. If you are just right, add another 10 degrees of flap, retrim as necessary, and continue on. And one other important thing—use your power, as you need it. That is why the throttle moves. Small power changes, introduced at the exact time they are needed, do much to smooth out the approach. Don't be timid with the power.

Let's break here for a moment and try to answer one of the most frequently asked questions concerning the traffic pattern. What do you do if you are too high or too low? The answer is about 50% common sense, 48% experience, and 2% instruction. Since there are so many possible combinations of errors, here are some general pointers.

If you are gliding in with constant power, constant airspeed, and flaps full down, and you are getting low, the first thing to do would be to add some power. Chances are this solution would take care of the problem. Fly the aircraft up to the point where you re-intercept the glide path, reduce the power back to approach setting, and continue on to land. If a little power doesn't work, try a lot. Use it as you need it. I find it much preferable to add a lot of power and land on the runway than to maintain a beautiful constant glide with the airspeed and attitude constant and land in the mud. It's hard to taxi that way.

Fig. 4-5. *This pilot has waited too long to turn from base leg to final approach, a very common error. This is a prime area for an approach-to-landing stall.*

Most people don't have too much trouble deciding what to do when they are too low. I guess it's the ground coming up at them that shakes them into action. And, if you find yourself too high, there are many things you can do to alleviate the situation. However, when some students are too high, it's an entirely different matter. They just sit there. I guess they figure it will come down sometime. It will, probably in an orchard or subdivision. So you see, there is as much reason to act if your aircraft is about to overshoot as there is if you are too low. If you don't put it on the runway, the results are usually the same—trouble.

If you are too high and have not yet put down all of your flaps, add some more flaps to increase your rate of descent. Or you can opt to remove some, or even all, of your power. Maybe you need to do both, add flaps and take off some power. Yet another option, used less and less these days, is a *slip*. A moderate forward slip will increase your sink rate a great deal and turn a possible overshoot into a workable approach. I teach my students all three methods and have them use them in the same order just talked about. If they are high, they put down some flaps. If they are still too high, they reduce power. If that doesn't work, they slip it. Of course, if you see you are still going to overshoot, nothing replaces the go-around, try-it-again method. In any case, don't just sit there and wait to see what happens; take action.

Back on base leg, we left ourselves in the key position with 20 degrees of flap and normal approach speed. The next step is to start the turn to final early enough so you can make a shallow turn from base leg to final approach, keeping the runway in sight throughout the turn. The most common error in this area is a pilot who starts the turn too early and makes the turn so shallow that they actually angle toward the runway instead of squaring the corner to complete the rectangular pattern. This results in actually arriving at the runway threshold at an angle to the runway instead of being aligned with it. All this does is further complicate an already difficult situation for the student pilot.

COTTER-PIN APPROACH

The opposite of this procedure occurs when the pilot starts his turn to final approach too late, tries to utilize the normal bank he would have used had he started the turn at the correct place, and proceeds to complete what I call the *cotter-pin approach*. He overshoots the runway centerline and then has to steepen the bank beyond safe limits in order to turn back to intercept the runway centerline. Viewed from above, it looks like a cotter pin. And it can become very dangerous very quickly.

A seemingly harmless item like turning from base leg to final approach too late catches the unwary pilot in a situation of deteriorating control command, increasing bank and stall speed, and decreasing airspeed. If I ever wanted a recipe for a potential disaster, this would be it. And the innocent thinking of the unknowing pilot goes something like this:

"Oops, I turned too late—guess I'll just steepen my bank a little and get back on final to the runway. I remember my instructor told me as long as I didn't exceed 30 degrees of bank in the pattern, I'd be okay."

"Wait a minute, this still isn't getting it. I'm not turning fast enough to line up. I bet if I add just a tad of inside rudder (in the direction of the turn), my rate of turn will increase and I'll be okay. I certainly don't want anyone to see me go around. They say it's a sign of a poor pilot." (Who are they? Does anyone know? I've never been able to figure it out.)

"Oh, oh. My addition of rudder has made my bank begin to increase beyond 30 degrees. They say that's not good. I'll just add a little bit of opposite aileron to counteract the rudder."

"Darn, still not getting turned quickly enough. I know. I remember from ground school that if I want to increase the rate of turn, all I have to do is add backpressure. Here goes. DAMN!" The airplane stalls, spins, and probably crashes. And it can happen about that fast.

The pilot in the above scenario has been tempted, trapped, and most likely killed by a combination of hangar talk, misunderstanding of flight dynamics, and poor association and correlation of aviation procedures. He has used some right ideas in the wrong places, ignored the bank-to-stall ratio and allowed some interesting, but incorrect, precepts to become fact.

Look at the procedure he used. He turned too late—added to his problems by adding inside rudder with opposite aileron (a slip), and then added the final piece to the recipe by pulling some more backpressure.

Think about this. The airplane is banked at 30 degrees—airspeed at about 70 knots. A cross-controlled slip is added which further reduces airspeed and increases stall speed. Then cross-controlled, at a slow airspeed, the pilot adds some backpressure. This is one of the best ways on earth to describe the entry to a snap roll. An aerobatic maneuver of immense fun and requiring a lot of skill. It is not designed to be performed by fledgling pilots on final approach! It will kill you! But it continues to happen. And I cannot quite figure out why.

The solution to this very real and present problem is to plan ahead and start your turn from base to final early enough so you can utilize the ground track procedures you learned earlier. You'll wind up with an ever-decreasing bank as you reach the final approach course. It's also easier than doing S-turns on final and hunting for the runway centerline.

On final, keep aligned with the runway centerline and add the remainder of your flaps as you need them. Maintain a constant airspeed and attitude. Power is reduced as you no longer need it, until, if all goes well, you reduce the power to idle in the landing flare.

If you have the approach pattern down pat, the landing should come rather easily. As I said before, it is just a transition from the normal approach to the landing attitude. This transition usually begins at about 25 to 30 feet above the runway with you *slowly* increasing the backpressure as you continue to sink. If you have too much airspeed or pull back too rapidly, you might actually climb a little. You don't want to do that, so increase the backpressure gradually. In a good landing, you almost think of it as trying to hold the aircraft off the ground through increased backpressure. You are transitioning the aircraft from a nose-low approach attitude through level flight to a slightly nose-high attitude at touchdown. If everything is right, you reach the point where you run out of back pressure just an instant before the touchdown occurs on the main wheels. Since you are nose high, the nose gear will still be up off the runway at touchdown. As your

Fig. 4-6. *A normal landing with full stall, full flaps, and full back stick.*

speed decreases following touchdown, allow the nose gear to lower to contact the runway and thereby provide you with more positive directional control (FIG. 4-6).

The landing is far from complete merely because you are on the ground. Many aircraft accidents occur during the rollout following touchdown. Keep your eyes outside of the cockpit, making sure you are maintaining runway centerline during the slowdown process. You can turn off the carb heat and bring the flaps up after you have cleared the active runway and are positive everything is under control. More than one beautiful approach and landing has been spoiled by the pilot fumbling with something inside the cockpit too soon after touchdown, only to be rudely awakened by the sound of runway lights shattering as the aircraft wandered off the runway. Remember that the control effectiveness steadily decreases as your speed decreases—just the opposite of takeoff. The slower you go, the more control movement it takes to get the job done. Use care but don't overcontrol in this phase either. Use whatever it takes—no more and no less.

Most of the errors common to landings, outside of a poor traffic pattern, have to do with your sight reference points. Nobody can show you exactly where to look, but you can gain some insight in where *not* to look:

- Do not look down and to the side while landing. Speed blurs vision, and you will not be able to tell if you are five feet or five inches above the runway by looking down and to the side.

- Don't look too close in front of the aircraft during the landing flare. Pilots who do so have a tendency to flare high, stall, and drop in.

- Don't look too far ahead of your aircraft during the landing flare. Due again to depth perception, this causes many pilots to run the aircraft into the ground with little or no flare.

- Once you are in the flare, don't look inside the cockpit for any reason. You've got to see where you are going.

Now that you know where not to look, where do you look? During the short final and flare, look approximately as far ahead of your aircraft as you would if you were in a car traveling at the same rate of speed. If you don't drive, I guess it's trial and error.

One other very common problem associated with learning to land is learning to pull the yoke straight back. I have had students who would begin to flare perfectly, and then at about 10 feet up, turn and try to take it to the tie-down area. The problem lies in flare technique. In most aircraft, the flare requires the coordinated use of the hand, wrist, elbow, and shoulder. If you try to flare using only your hand, wrist, and elbow, the tendency is to twist the elbow up and out, causing the ailerons to be deflected to the right. It also causes a right turn tendency that can spoil an otherwise decent approach and flare. Conversely, if you pull your elbow in toward your body as you flare, you will experience a left turning tendency. So watch for these common errors and if you have trouble in this area, sit in the aircraft and practice pulling the yoke straight back.

CROSSWIND LANDING

It seems that not too many people pay much attention to slips anymore. I guess the advent of flaps and spoilers has caused the need for the slip to be pushed to the rear of the list of things some pilots believe they ought to learn. However, they still have a very real place in the mind of the complete pilot. The FAA has only recently reintroduced the forward slip to a landing as a required maneuver for the Private Pilot checkride.

Some say, "If you have flaps, you don't need to slip, and anyway, it's unsafe to slip with your flaps down." To this I politely say, "Nonsense." Unless your particular aircraft is placarded against slipping with flaps, go ahead. And even in the aircraft that are placarded concerning slipping with flaps, most don't prohibit them. The placards usually say, "Avoid slips with flaps extended." Many aircraft flight manuals don't mention the fact one way or the other. If your aircraft is prohibited from slipping with flaps, don't. Otherwise, why not? It just might get you into a field you might have otherwise overshot. It can be especially true in an emergency situation. I, for one, am not going to sit and wait patiently for the trees to gather me up into useless aluminum and flesh if I can slip it down on a nice, smooth, green field. For those who are not going to slip no matter what, skip this part. The rest of us are going to learn another step toward becoming a complete pilot.

Generally, there are two varieties of the slip: the *forward slip* and the *side slip*. Although the two are very much alike in the manner in which they are executed, the forward slip usually requires much larger doses of control input. The one facet in which both are alike is the fact that you must have your controls crossed. Crossed controls

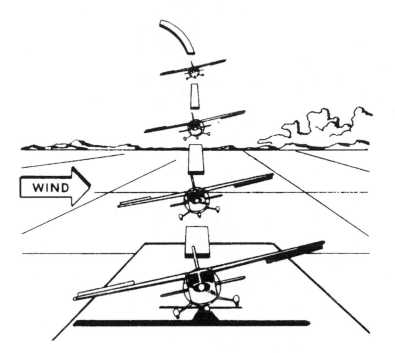

Fig. 4-7. *The side slip approach to a crosswind landing.*

means you hold your ailerons in one direction and rudder in the other-something most pilots find rather uncomfortable until they have executed many slips.

If you have any wind, the slip should be done into the wind. (See FIG. 4-7.) If the wind is from your left, the left wing is the one put down. But, if all you do is put down the left wing, what will happen? You'll turn left. To keep the aircraft from turning, add opposite rudder. Now you are cross-controlled—slipping.

To put these slips to work for you, you need to know what each one is for and what the desired results should be. The forward slip is used mainly for altitude loss. The sideslip is used to align the aircraft with the runway and to touch down in a crosswind.

The Forward Slip

Let's set up a situation where both types of slips are used on one final approach and landing. Throw in a crosswind to complete the setup. You have turned from base leg to final approach and find yourself very high. You are landing on runway 18 and the wind is from 120 degrees at 10 knots. You already have full flaps and power at idle, yet you are still going to overshoot. Now what? If you are very high, you will have to go around; however, if you are only a little high, the forward slip just might be the ticket (FIG. 4-8).

Lower your wing into the wind as you apply opposite rudder. In this case, you need to lower your left wing and add right rudder. You want the nose to swing past direct runway alignment, say 15 to 20 degrees to the right. You will still be tracking straight down the runway, but your nose will be pointing at a heading of about 200 degrees or so. This is the signature of a forward slip: Your heading changes while your ground track remains the same.

Now comes the important part: Don't let your airspeed build up much over approach speed. If you do, all the gain of the slip will be lost because once you get down, you will have to bleed off the airspeed and you will float much farther, negating the advantage you gained from slipping.

Continue to slip until you are down to the point where you reintercept your glide path. Then, come out of the slip and proceed with your approach. You get out of the slip by reversing your aileron and rudder, bringing your aircraft back to level flight with the ailerons, and using left rudder to swing the nose back to direct runway alignment. Most likely, since you have a crosswind, you should continue to bring your nose past direct runway alignment into a crab condition to take care of the crosswind ground track. Now, you have lost the unwanted altitude, are still tracking straight down the runway, and will be ready to complete the crosswind landing.

The Sideslip

The *crosswind landing* is accomplished using the sideslip. In this slip, the longitudinal axis of your aircraft remains parallel to the flight path. Since you don't want to land in a crab (unless you are fond of tires screeching and the sound of aluminum bending), you

Fig. 4-8. *A forward slip as viewed from the cockpit. The wing is down into the wind, and the aircraft is actually pointing at the terminal buildings while the ground track remains straight down the runway.*

Fig. 4-9. *A sideslip as viewed from the cockpit. The wing is still down into the wind, but the aircraft is now aligned with its longitudinal axis parallel to the runway. This is the correct way to land in a crosswind from the right.*

need a way to get your aircraft straight down the runway before touchdown. The sideslip is, I believe, the best answer.

In the sideslip, lower your wing into the wind and add opposite rudder as you did in the forward slip (FIG. 4-9). Add only enough rudder to maintain your track straight down the runway. You sort of lean into the wind like you used to do when you rode your bike in a strong crosswind. You still go straight, but are banked to correct for the crosswind. It probably will take a little practice before you become proficient and comfortable with sideslips and remember to use the controls as you need them. You can't just put the wing down a certain number of degrees and leave it there because the wind will probably increase and decrease in intensity. So use whatever it takes to hold your runway alignment.

The crosswind landing is accomplished exactly as any other landing, except you keep your sideslip throughout the approach, flare, and touchdown (FIGS. 4-10 through 4-12). Remember, after touchdown, you're not through yet. Maintain your upwind wing down into the wind and use your rudder as necessary to maintain runway heading. As you slow down, you will want to add more aileron into the wind because the controls become less and less effective as your speed slows. In fact, as you reach your slowest speed, you will want to have full aileron into the wind to correct for the crosswind, just as if you were taxiing. By the way, that is taxiing.

Fig. 4-10. *Crosswind landing approach. Wing down into the wind with enough rudder to hold the aircraft straight down the runway.*

Fig. 4-11. *Just prior to touchdown in a left crosswind. Wing is still down and rudder is used as needed to maintain alignment.*

SOFT-FIELD TAKEOFF AND LANDING

Soft-field takeoff and landing techniques are useful to all pilots. They are used to get into or out of fields made soft by mud, snow, tall grass, or even just wet grass. And this technique is also very good for fields where the surface is rough. In fact, soft-field procedures are used almost anytime you are not using a smooth, hard-surfaced runway. The primary concern is to prevent the aircraft from nosing over in the soft substance during the take-off run or the landing roll.

Soft-Field Takeoff

For a soft-field takeoff (FIG. 4-13), your flaps should be set as per the manufacturer's recommended flap setting if your aircraft is so equipped. (Many low-wing aircraft manufacturers recommend using no flaps since the flaps on low-wing aircraft are very close to the ground and can easily be damaged by mud, rocks, etc.) Since the surface is soft, it is very important that once you start your aircraft moving, you keep it moving. Even taxiing should be done with full aft yoke to offset any noseover tendency.

Fig. 4-12. *Crosswind touchdown. Upwind wing down, rudder as required to maintain alignment; upwind wheel touches first.*

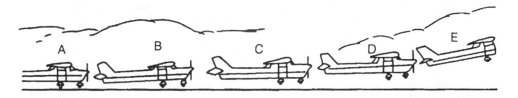

Fig. 4-13. *Soft-field takeoff: A) Smoothly apply full power and raise nosewheel clear of impeding substance; B) Lift off at slower than normal airspeed; C) Gently level off; D) Remain in ground effect until reaching V_x or V_y, depending on possible obstacles; and E) Upon reaching desired speed, gently pitch up and climb out normally.*

The pilot's first concern during the soft-field takeoff is to transfer the weight from the wheels to the wings as rapidly as possible. The flaps help do this, but at the start of the takeoff run, you should have the yoke full back to help lift the nosewheel off of the ground and reduce the drag created by the soft field. Since this technique gives you a much higher angle of attack than other takeoffs you have been used to, once the nose begins to come up, you must be prepared to relax a bit of the back pressure so you don't wind up smacking the tail into the ground. That is very easy to do, and the loud bang the tail makes as it strikes the runway usually scares a pilot into shoving the stick forward, sometimes a bit too far. The nosewheel might then get swallowed up by the soft surface you are trying so hard to get out of.

Once you have the nose at the desired attitude and are rolling down the runway, keep the nose attitude constant until the aircraft is off of the ground. The liftoff should occur at a much slower-than-normal airspeed because of the high angle of attack. For this reason, once the aircraft becomes airborne, the angle of attack must slowly and smoothly be reduced to near-level flight attitude as the aircraft accelerates toward the normal climb speed. You should maintain the near-level attitude, in ground effect, until you reach your nominal climb speed (either V_x or V_y, depending on whether or not there is an obstruction to be cleared).

After accelerating in ground effect to the desired speed, a normal climbout should be initiated and the flaps brought up after you are sure you are maintaining a climb. Don't be in a hurry to bring up the flaps. They are providing lift, and if brought up too soon, they could cause a momentary sink that could lead to trouble.

One other important point concerning soft-field takeoffs, or any takeoffs for that matter, is to be sure you lift off and maintain a straight track down the runway. If you have a slight crosswind, resist the tendency to crab into the wind as soon as you lift off. If the wind were to die, you could settle back to the runway in a crabbed configuration, and the result could be one of the shortest flights on record. Use the wing-low sideslip method we discussed earlier for crosswind takeoffs and landings. Keep your aircraft in this sideslip while heading straight down the runway and until you are positive you are going to remain in the air.

Soft-Field Landing

The approach to the soft-field landing (FIG. 4-14) should be made at normal approach speed. The touchdown should be as slow as possible in order to minimize the noseover tendency. Unless you are in a low-wing aircraft and the manufacturer recommends no flap landings in a soft field, this usually means an approach and landing with full flaps. The touchdown should be with flaps as recommended, full stall, and full back stick. A small amount of power can be used in the flare to help bring the nose up and provide the momentum necessary to prevent a noseover.

Keep the yoke in the full-back position during the rollout and be ready to use power to help you through any really soft spots like snow drifts or deep mud. When you are safely on firm ground and slow enough to taxi, you can take the carb heat off and bring

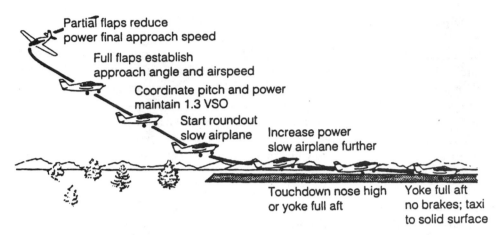

Fig. 4-14. *Soft-field landing.*

up your flaps. Use caution as you taxi because you don't have it made until you have the aircraft safely tied down.

Note the following example from NTSB file # 3-3611. On December 18, 1976, a Cessna 170 with a 38-year-old commercial pilot attempted a landing at an airport near Ashland, Montana. The airport and the surrounding terrain were snow covered. Phase of flight: landing roll. Probable cause: unsuitable terrain. Remarks: landed on snow-covered strip, rolled into deeper snow, and swerved, noseover.

An important item to remember during the conduct of soft-field operations is that it is nearly impossible to judge when snow or mud get deeper. All one can see is the top of the surface and there is no way of knowing where the drifts or sinkholes begin. Soft-field operations should be handled with great care and forethought, not with haphazard preparation or spontaneous whims.

SHORT-FIELD TAKEOFF AND LANDING

The short-field takeoff and landing is used for one of two purposes: 1)To get you into or out of fields that are actually physically short. Or, 2) To take off or land in fields which have some type of obstruction that reduces their effective useful length. In either case, the short-field technique should be used to take off or land safely.

Short-Field Takeoff

The very first consideration for anyone contemplating a short-field takeoff is to find the distance it will take for your aircraft to accelerate, lift off, and clear any obstacles. The first place for you to look would be in your aircraft's flight manual in the performance section. Here you will find a chart disclosing takeoff distance information allowing for such conditions as type of runway surface, temperature, and wind. The figures will show you the required distance both to lift off and to clear a 50 foot-high obstacle for a given set of conditions. And remember, the distance in the flight manual is set by test pilots flying new aircraft with new engines in top condition and under controlled circumstances. In other words, these distances should be thought of as probably the absolute minimum and are not a guarantee of the distance it might take you to get airborne and over the obstacle.

Assuming you are taking off from a short field (FIG. 4-15) or a field with an obstruction, you should strive to learn to fly your aircraft by airspeed and attitude control rather than letting your instincts take over. If you are caught trying to clear an obstruction or take off before the aircraft is ready to fly, the results can be less than desirable. If you try to force your aircraft into the air before it is ready to fly, it might drop back onto the runway and actually lengthen the time needed to clear the obstruction.

When starting to learn the short-field takeoff procedures, a very good habit to get into is to use every available inch of runway as you line up to practice short-field takeoffs. Any runway left behind you might as well be in another country. It will be of no use to you.

Utilize the manufacturer's recommended flap setting if the aircraft is so equipped. Line up, using the entire runway, and smoothly, but firmly, add full power. Don't hold the brakes. Get going! Studies have shown that, unless you are in a turbine-powered aircraft, holding the brakes while running the engine up to maximum RPM before brake release does nothing to enhance the short-field takeoff distance. It might, in fact, hinder your acceleration and thereby lengthen your takeoff roll.

Smoothly apply full power and let the aircraft accelerate until you arrive at the best angle-of-climb airspeed (V_x). Keep the aircraft directly parallel to the centerline and let it seek its own pitch attitude. That is, don't force the nose down while you are accelerating because it will likely give you a negative angle of attack and increase the distance it takes you to arrive at V_x. When you arrive at V_x, rotate and maintain that airspeed until you have safely cleared any obstruction. After clearing the obstruction, the angle of attack should be reduced and the aircraft allowed to resume its normal climb speed (V_y). The flaps should be left alone until you have safely cleared any obstacle, accelerated to V_y and then they should be brought up very slowly.

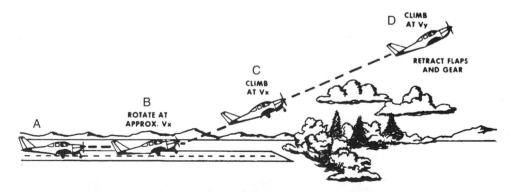

Fig. 4-15. *Short-field takeoff: A) Smoothly apply full power; B) Rotate at V_x airspeed; C) Maintain V_x airspeed until clear of obstruction; and D) lower pitch to V_y and climb out normally.*

While on the subject of short-field takeoff technique, I would like to mention a very large difference of opinion between instructors, pilots in general, and the FAA. Some instructors tell you to rotate before you get to V_x on a short-field takeoff. They say that by the time you rotate, you will arrive at V_x and everything will work out better. They say it is almost impossible to rotate and maintain V_x. I say they are wrong. The FAA's own handbook says they are wrong. Many aircraft flight manuals say they are wrong. If rotating at and maintaining V_x is beyond their capability, then let them rotate early, but we who have the skill to rotate and maintain V_x will beat them over the obstacle every time. Remember, the idea is to go over the tree, not through it.

Short-Field Landing

Short-field landings (FIG. 4-16) should be practiced assuming a 50-foot obstacle exists on the approach end of the runway. The approach to a short-field landing should be made with power as needed and at a speed no slower than 1.3 V_{so} (1.3 times the power-off stall speed with your gear and flaps down). A competent pilot should be able to execute the approach as if it were "descending slow flight."

Your final approach should be longer than normal because it's on final that you will find the key to establishing your short-field technique. This key is one of obtaining a clear mental and visual picture of a straight line from your position on final, over the obstruction to the point of intended touchdown. Once you have set up this mental picture, attain the desired airspeed and control the descent using coordinated power, flaps, and proper pitch attitude. If it works out right, your power should be slowly reduced until it reaches idle in the landing flare. The point I'm trying to make is that a short-field approach is a *powered* approach. As you no longer need the power, you slowly get rid of it.

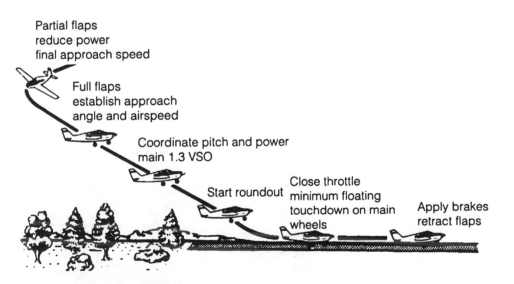

Fig. 4-16. *Short-field landing.*

While on final approach, if you see that you are going to be low, use additional power to reestablish your glide path. If you find yourself a bit high, merely reduce your power a little in order to attain a higher rate of descent until you arrive back at your proper approach angle and proceed normally. If you are way off either way, too low or too high, you would probably be wise to go around and try it again.

The point I'm attempting to hammer home is that your glide path is adjustable. It's up to you. By merely adding or reducing power you can increase or decrease your rate of descent and adjust your approach path accordingly. You are actually slow flying your aircraft down final approach adjusting your altitude with your power and your airspeed with your pitch.

The touchdown should come at minimum controllable airspeed, with power at idle, and with little or no float. Your flaps should be retracted as soon as possible after touchdown to place all of the weight on the wheels and aid in braking.

Aside from improper pitch and power control, there are two very common and potentially dangerous errors in the execution of the short-field approach and landing. One is the tendency to lower the nose after clearing the obstruction, which results in an increase in airspeed and causes the aircraft to float, using more runway than you would have used if the attitude had been held constant all the way to the flare. As you might imagine, this could lead to an overshoot that could be as deadly as landing short.

The second error is reducing the power to idle as you cross over the obstruction. This procedure can be very dangerous because if you are approaching at a very low airspeed, the sudden loss of thrust can lead to an immediate stall. Keep your attitude constant down to the flare, and use your power as you need it. Do not reduce your power to idle until you are in the landing flare and are very close to the runway.

Strong or gusty winds can cause the short-field approach to become a little treacherous. It is wise to carry a little more power and airspeed in these conditions. A strong wind will slow the groundspeed of your aircraft so a little more airspeed will bring about the desired results with an added amount of safety. Gusty winds are even more troublesome. Again, the power and airspeed should be a little higher than normal. On the approach, if the wind you are riding suddenly dies, the aircraft is likely to sink rapidly and leave you little or no time to recover. More than one pilot has been the victim of this bit of treachery. Remember, there is a huge difference between a short-field landing and landing short of the field.

If you have trouble with normal short-field approaches, return to approaches at 1.3 V_{so} and gradually work down to slower speeds. Don't try to force proficiency from yourself before you are ready to handle it. Short-field techniques require a high degree of reflex and feel for the aircraft (FIG. 4-17). These are things that cannot be taught by the best instructor. They must be acquired by trial, practice, and time.

GO-AROUND

One of the most important facts a truly competent pilot carries in his mind is that not all approaches can be successfully completed. Perhaps you have misjudged the wind or are too high or too low to comfortably complete the approach. Or maybe someone has pulled

Fig. 4-17. *The proper landing attitude for normal, short, or soft fields. Full stall, full flaps, full back stick.*

out onto the runway in front of you. In any case, the prudent thing to do is to go around and try it again.

Some pilots feel a go-around is a sign of poor pilot technique or a lack of piloting skills or just plain embarrassing. Wrong. A go-around, done because it is needed, shows the pilot is thinking safety, has a good grasp of the situation, and would rather let the lesser pilots razz him a little than compromise safe flight technique.

Generally speaking, a go-around is merely the transition from approach configuration back to normal climb attitude. A little dose of common sense, along with some dual, should help you realize the ease and importance of this action. If you have misjudged your approach, the go-around can be done as you continue your track straight down the runway. If someone has pulled onto the runway in front of you, you should move off to the side in order to keep the other traffic in sight as you continue your go-around. Since you will probably be in the left seat, this would mean moving off to your right, just to the side of the runway, not into the next county.

Let's say you're on short final with full flaps and low power and you have to execute a go-around. Proceed with these steps:

- First, add full power as you begin the transition through level flight to a climb attitude.

- Second, take off the carb heat in order to develop full power.
- Third, bring the flaps up to the manufacturer's recommended go-around setting. This decreases your drag and allows you to accelerate better.

After completing all of these steps, make sure you are at least at V_x airspeed holding your own, or better still, climbing. When you are sure the airspeed, heading, and altitude are stable, retract the remainder of your flaps, allow the airspeed to accelerate to V_y, and continue a normal climbout.

In the event the situation requires a side step to avoid another aircraft, the go-around would be completed as shown above except you would have to make a shallow turn. Unless the situation requires your immediate action, I believe it is better to get powered up and cleaned up before attempting the turn. Then, you will not get involved in a low airspeed, low power, flaps-down bank that can prove to be troublesome, even for a very experienced pilot.

There is one other problem that might necessitate a go-around—a bounced landing. If you haven't done any yet, you will. Most of the time, the bounce is slight and you can recover by adding a little power, lowering the nose to level flight attitude, and then flaring again—most of the time.

But if you really bounce it by dropping in from about 15 feet and spreading out your spring-steel gear to the point it groans, snaps back into position, and sends you about 30 feet back toward the sky, you just might want to consider a go-around. Let's face it, you're already up there, so you might as well go around. When this happens, stay calm and smoothly add full power. Lower your nose to level flight attitude and accelerate to climb airspeed as you slowly retract your flaps to the manufacturer's recommended go-around setting. You will probably experience a little sink as you bring the flaps up, but don't pull the nose up. Leave it at about level flight attitude so you can gain speed more quickly. Complete the go-around by heading straight down the runway, and climb out normally. I believe you will find the next landing to be much better.

Note the following example from NTSB file # 3-3912. On December 9, 1976, near Spokane, Washington, a student pilot with 16 total hours was practicing landings when an accident occurred that caused substantial damage to his American AVCO AA-1. The type of accident is listed as a hard landing and gear collapse. Probable causes: improper level-off and improper recovery from bounced landing. Factors: touch-and-go landing, overload failure, nose gear collapse, nosed over. Minor injuries.

Touch-and-Go Takeoff and Landing

Before leaving takeoffs and landings altogether, I would like to offer my input on a debatable topic. I feel that touch-and-go takeoffs and landings have no place in the practice of flight maneuvers. They are a waste of the student's time and money and the learning process. Why? Because you never get to experience the vast changes in control response that comes from the airspeed changes that occur in every full stop landing.

Let's assume the approach speed of your aircraft is 75 knots and the stall speed in landing configuration is 40 knots. During the execution of a touch-and-go landing you will only experience the change in control forces that come from the difference between 75 knots and 40 knots, a 35-knot variance. But, if you come to a full stop after the landing, you will experience a 75-knot variance in airspeed and the resultant variance in control response. You are missing out on one of the most important elements of flight—learning the feel of the controls as the speed increases and decreases. Changing conditions, changing winds, and changing speeds are what flying is all about. You just don't get the full benefit of these changing conditions that are so important to your learning process if you shoot a lot of touch-and-go takeoffs and landings (FIG. 4-18).

If you follow aviation for many years, as I have, it begins to dawn on a person that accidents don't just happen—they are the result of something missing in a pilot's training or in their attitude. Such is the case, I believe, with touch-and-go takeoffs and landings. If you were to follow the NTSB files for a bit, you would begin to see a pattern develop to landing accidents and incidents. Namely, most of them do not occur just after touchdown when the aircraft is going the fastest. The accidents seem to occur as the aircraft slows down to below touch-and-go speed, a speed that many pilots haven't spent much time exploring because they were too busy squeezing in one more bounce and go during their training and never spent much time assessing the control needs at the slow end of the spectrum. The result is a rash of unpleasant surprises that come to visit the unwary pilot sometime during their early crosswind exposure after they attain their license.

In my career, I have personally witnessed two landing accidents. In each case, the aircraft had just landed in a medium crosswind from its left and was rolling out to com-

Fig. 4-18. *Everybody wants to park closest to the terminal!*

plete the landing. Each aircraft was traveling less than 40 knots when it swerved, dipped a wing, struck the ground with the down wing, and rolled over.

In one case, I was on short final directly behind the action. Boy, what a sight. The aircraft swerved a bit, corrected, and then weathervaned into the wind as it dove for the runway's left edge. I mean, this clown took it so far from the runway before he flipped it over that I landed on the runway and taxied over and gave him a hand getting out.

The point is, in both cases these pilots lost it after landing in a pretty good fashion. They lost it when their control effectiveness reduced to a point where they had not spent much time—below 40 knots. It got them because they were not used to the feel required to elicit control response at these low airspeeds. Had they spent more time training on takeoff and landings to a full stop, particularly in a crosswind situation, these accidents most likely would not have occurred.

5
Advanced Maneuvers

IN THIS CHAPTER, WE WILL CAREFULLY DISCUSS AND ANALYZE THE so-called advanced maneuvers, including the chandelle, the lazy 8, and the pylon 8. These three maneuvers, although differing greatly, have a common goal: coordination. To improve your coordination is to improve your flying. To improve your flying is to increase your safety margin to the point where each flight leaves very little doubt of a successful completion. Coordination and safety go hand in hand.

These advanced maneuvers might offer you a bit more of a challenge than you have been used to. But the maneuvers are worthy in themselves since they cause you to expand your skills and those of your aircraft. In other words, these maneuvers will take pilots out of their comfort zone and extend the need for insight into just how and why the maneuvers work. The end result should be a better understanding of flight and a higher level of proficiency for the pilot.

Many pilots seem to fear these advanced maneuvers for one reason or another. In my nearly three decades as a flight instructor, I have noticed that the reason for this fear is usually ignorance. I don't necessarily mean the pilots were stupid, but they were ignorant of the makeup of a given maneuver. They didn't understand how a maneuver is made up of many smaller components that are grouped together to construct a new maneuver. They tend to forget that *all* maneuvers are comprised of straight-and-level, climbs, turns, and glides.

CHANDELLE

The *chandelle* is an advanced training maneuver that requires a great deal of coordination and preplanning in order to be successfully completed. In addition to its training status, the chandelle also has some very practical applications in everyday flight. For instance, because the chandelle is basically a 180-degree climbing turn, you might use it to turn up and away from another aircraft, up and away from adverse weather, or, possibly, to reverse your direction in a canyon you discover you cannot outclimb. It has more far-reaching possibilities than just another maneuver to learn to please the examiner at checkride time.

To initiate your practice of a chandelle, climb to at least 1,500 feet above the ground because the maneuver requires the completion to be near stall speed, and minimum stall recovery altitude is 1,500 feet AGL. Line up crosswind, because all turns should be into the wind in order to remain in the practice area. As far as the aircraft is concerned, it really doesn't matter which direction you turn. You won't gain any more altitude one way or the other. In addition, the aircraft should be at, or below, V_a (maneuvering speed) to initiate the maneuver. Depending on the aircraft, you might have to dive or slow down slightly in order to attain this recommended entry speed.

Pick a minimum of three reference points in order to complete the chandelle using outside visual references: one directly ahead, one at the 90-degree point, and one at the 180-degree point where the maneuver will be completed (FIG. 5-1).

When you are at least 1,500 feet above the ground, at V_a, set up crosswind, and with your three reference points, you are ready to begin the maneuver. Roll into a moderate bank (about 30 degrees) and begin to smoothly increase your pitch. Maintain your 30-degree bank and steadily increase your pitch until you reach the highest point of pitch at the

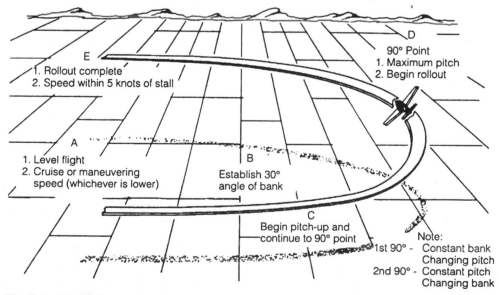

Fig. 5-1. *Chandelle.*

90-degree point of the turn. Somewhere around the 45-degree point you will have to begin to add a bit of right rudder to correct the left-turning tendency caused by torque and P-factor. This application of torque correction during turns in both directions helps in making the turn rates constant, which is very important in the overall coordination process.

Upon reaching the 90-degree point, maintain a constant pitch attitude and begin to roll out of the 30-degree bank proportional to your rate of turn. Continue to correct for the effects of torque as you keep your pitch attitude constant through the second 90 degrees of the maneuver. Your rollout should be timed so that as you arrive at the 180-degree point, the wings will just be coming level and your airspeed will be just above a stall.

Note: Any pitch change after reaching the 90-degree point is evidence of improper pitch control and planning.

To maintain a constant pitch attitude while you are turning from the 90-degree to the 180-degree point, bring the elevator control back slightly to compensate for the loss of airspeed and the resultant control ineffectiveness. You see, as your airspeed slows, you will need more elevator input to maintain the desired pitch attitude. And a great deal of the success of this maneuver depends on maintaining constant pitch during the 90-through-180-degree portion.

The most common error made in executing the chandelle, other than the obvious pitch and bank errors, is a tendency for many pilots to hurry the maneuver. Do not hurry this maneuver. Chandelles done as slowly and as smoothly as possible are usually the best. This results in the greatest gain of altitude, and more importantly, helps you feel the vast difference in control effectiveness as your aircraft slows from maneuvering speed down to just above a stall. Also, watch for that torque and P-factor correction mentioned previously. Proper coordination goes a long way in helping your chandelles become meaningful and precise.

LAZY 8

The *lazy 8* is another advanced training maneuver requiring preplanning, coordination, and timing. One interesting aspect of this maneuver is that there are almost as many ways of performing it as there are pilots. However, two common denominators emerge: The lazy 8 is a superb training maneuver for everyone. And the lazy 8 can be unequalled in frustration for those who do not understand the various aspects of it. For those unfortunate pilots who draw an instructor who doesn't personally understand a lazy 8, life can be pretty miserable.

The lazy 8 is really a fairly simple maneuver if you are well schooled in your basics and the use of outside visual references. The lazy 8 is a compound training maneuver combining just about all of the aspects involved in flight and garners its name from the fact that the nose of the aircraft will scribe an 8 laying on its side as the maneuver unfolds. If you had a pencil attached to the spinner on the nose of your aircraft, the climbs, turns and descents of this constantly changing ballet would draw an 8 lying on its side. In the performance of a lazy 8, the controls are always moving; bank is never constant, heading is never constant, and altitude is never constant. The lazy 8 is a graceful, ever changing, symmetrical dance through the sky.

Chapter Five

And one more important note: The lazy 8 is the only flight maneuver I am familiar with that cannot be done by rote. If the pilot performing the lazy 8 doesn't really understand them, it is abundantly evident from the beginning. I have had applicants for a Commercial license try and sneak an ugly lazy 8 past me on more than one occasion. It is sort of like trying to hide a fire in the dark.

To begin your practice of the lazy 8, climb to at least 1,500 feet and align your aircraft crosswind as you did for the chandelle. For the same reasons as the chandelle, all turns are made into the wind (FIG. 5-2).

Five outside visual reference points are needed to help you complete a lazy 8. These reference points have to be on the horizon directly in front of your aircraft and at each 45 degrees of turn. In other words, the five reference points begin with the first directly ahead and one at each of the 45-, 90-, 135-, and 180-degree points.

A well-executed lazy 8 begins and is completed at the same altitude and airspeed. For this reason, power selection is an important factor. You don't want too much power, which can cause you to gain more altitude than you can comfortably lose. On the other hand, too little power can cause too little altitude gain and destroys the symmetry of the lazy 8. So choose a power setting that allows you to begin at or slightly below maneuvering speed and you should be very close.

Unlike the chandelle, where you began the roll and then initiated the pitch, the lazy 8 calls for a simultaneous initiation of both bank and pitch. And the bank and pitch are begun slowly. Remember the name—*lazy 8*. It's not an accelerated 8 or an abrupt 8.

Slowly and smoothly begin the pitch and bank simultaneously. The pitch and bank are continuously and slowly increased until your aircraft arrives at the 45-degree point of your turn as the highest pitch is reached and your bank is arriving at about 15 degrees. At this

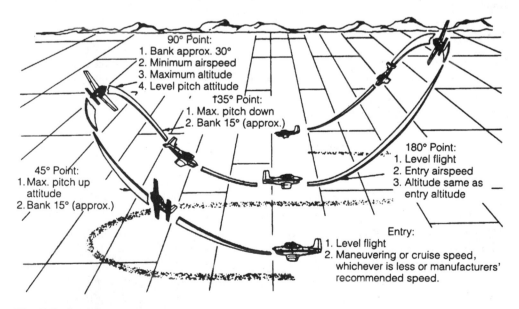

Fig. 5-2. *Lazy 8.*

time, due to very slow airspeed and high angle of attack, continued application of bank can cause your aircraft to lose some of its vertical lift. The aircraft will then fly down through the horizon at about the 90-degree point of the turn as your bank reaches the 30-degree point.

As you pass through the 90-degree point, some backpressure is released and your bank is reversed so you arrive at the 135-degree point with your bank back to 15 degrees again and your nose as much below the horizon as it was above it at the 45-degree point. The purpose of the nose being equidistant above and below the horizon is to help ensure maximum symmetry during the maneuver.

From the 135-degree point, continue to slowly reduce the bank and adjust your pitch so that you arrive at the 180-degree point just as the wings come level and your airspeed and altitude return to their initial starting point. To complete the lazy 8, follow the same procedure in the opposite direction. Remember, it takes two 180-degree turns to properly complete a lazy 8.

Your timing and coordination are very important to the symmetry of the lazy 8. At no time during the maneuver should your controls be held constant. The pitch and bank are constantly changing during the climbing and descending turns, and corrections for torque and P-factor are needed during the climbing portions of the maneuver. This leads to a cross-controlled situation in the climbing turn to the left since the bank is continuously increasing as you turn in this direction. If right rudder is added to aid in overcoming torque (or no correction is made), then a situation is set up where you are either cross-controlled and coordinated or in a slipping turn.

There are three distinct errors common while learning to execute a lazy 8. The most common is to hurry the maneuver. As I mentioned previously, it is *not* called an accelerated 8; it should be done as slowly and as smoothly as possible for the best outcome.

The second most common error is for the longitudinal axis of the aircraft to pass through the horizon either too early or too late. It should fly through the horizon at exactly the 90-degree point. If the longitudinal axis passes through the horizon too early, you will usually complete the maneuver at an altitude much lower than the altitude at which you entered. This is because passing through the horizon early allows the aircraft more time to descend and it winds up using this time to descend more than it climbed, thus destroying the symmetry of the maneuver. Conversely, if the longitudinal axis passes through the horizon late, the maneuver will be completed at an altitude higher than the original. The reason for this is the exact opposite of passing through the horizon too early. The aircraft will not have as much time to descend, and this causes the maneuver to end up at an altitude higher than the one from which you started. You can cheat and force your aircraft to return to the original altitude, but the symmetry is destroyed and you make little gain in understanding the precision involved in mastering this maneuver.

The third most common error concerns power selection. The correct power setting is essential if the lazy 8 is to be performed with any degree of symmetry. For example, the power setting you choose on a day when you are the only person on board with half tanks of fuel and an outside air temperature of 35°F will be quite different than if you are loaded with full fuel on a very hot day. The reasoning behind this involves your power-to-weight ratio. If your aircraft is light and the day is cool, you require less power to lift the weight. So if you use too much power, you gain more altitude than you can lose without exceeding your entry airspeed, causing you to wind up higher at the end of the ma-neuver than you were at the entry. On the other hand, if you select too little power for a given day, you will not climb enough to make your lazy 8 symmetrical. So before you initiate your lazy 8, give some serious thought to what power setting should be right for the conditions you have on this particular day. Don't try to use the same power settings day in and day out; it won't work. Remember what I told you earlier. The lazy 8 is the only maneuver I know of that cannot be done by rote. The lazy 8 is either performed correctly or not at all.

PYLON 8S

The *pylon 8* is the FAA's "new" maneuver that must be mastered for the Commercial checkride. I call it new because the FAA dropped this maneuver in 1975 but reinstated it in 1986. Consequently, a good many pilots do not have any idea how it works, what it's for, or how to perform it. Let's try to settle those points right now (FIG. 5-3).

The objective of the pylon 8 is to develop your ability to fly your aircraft while you are dividing your attention between the airplane's flight path and the ground reference points. In other words, the FAA wants you to be competent enough to manage your air-craft while your attention is diverted outside for any extended period of time.

The pylon 8 involves flying your aircraft in two circles about points on the ground, forming an 8 lying parallel with the ground, while you keep your aircraft's lateral axis (wingtip to wingtip) on the point. However, this is a ground reference maneuver as op-posed to a ground track maneuver because you do not make any correction for your ground track. What I mean by not having any ground track correction is that you will make absolutely no correction for wind during the execution of the pylon 8. You drift where you drift and concentrate on holding your wingtip on the pylon.

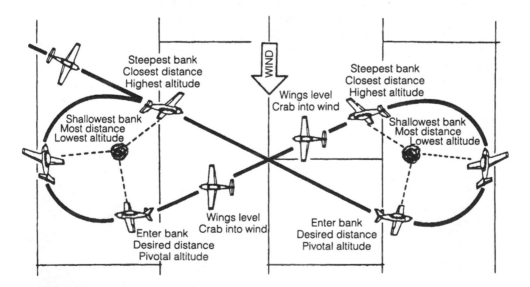

Fig. 5-3. *Pylon 8.*

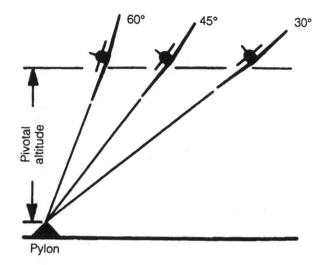

Fig. 5-4. *For a given pivotal altitude, your distance from the pylon increases as your bank decreases.*

However, during the pylon 8, if there is any wind blowing you will have a continuous change in your altitude. Your altitude will change as your groundspeed changes. Hence, a new term, "pivotal altitude," has been created (FIG. 5-4). Let me explain.

Pivotal altitude is actually a function of how fast you are traveling over the ground. In calm air, there is an altitude at which you could fly your aircraft about a point, keeping that point directly off the tip of your wing and your altitude would never change. This is *pivotal altitude*. But the wind is usually blowing, and this causes your groundspeed to change, so you have to change your altitude to make up for the groundspeed differential as you go from upwind to downwind. The faster your groundspeed, the higher your pivotal altitude; the slower your groundspeed, the lower your pivotal altitude.

To find your initial pivotal altitude, square your groundspeed and divide by 15. For instance, if your aircraft has a groundspeed of 100 knots, square it and you get 10,000 knots. Divide the 10,000 by 15 and your pivotal altitude is 666 feet. Now, add this to your terrain elevation and you have an indicated altitude to fly at to initiate the pylon 8.

Does it sound a bit difficult? Let's run through a set of pylon 8s to try to simplify it for you. To initiate your pylon 8s, first arrive at your pivotal altitude and pick two pylons in an open area. A line drawn between the pylons should be perpendicular to the wind, or crosswind, because you will want to enter the maneuver downwind at a 45-degree angle to, and in between, the pylons.

Okay, you have pivotal altitude attained, pylons picked, and you're heading downwind between the pylons at a 45-degree angle. As the pylon arrives at the wingtip reference, roll into your bank and place the pylon the same distance above or below your wingtip as the horizon would be in straight and level flight. Now, here is the key word that unlocks pylon 8s: *anticipate*. Anticipation of the groundspeed changing as you turn from downwind to upwind allows you to be ready to change your pivotal altitude as need arises (FIG. 5-5).

As you turn from your downwind heading, the groundspeed drops and your pylon appears to move forward as seen from its relationship to the wingtip reference. When this occurs, simply move the elevator control forward to maintain the pylon on its original position. When the pylon ceases moving, you are at a pivotal altitude for that particular groundspeed and are ready to anticipate the next wind and groundspeed change (FIG. 5-6).

Continue around the pylons, keeping in mind where the wind is in relation to your position, and anticipate moving the elevator in the same direction as the pylon moves. When you have turned about the first pylon to the point that the second pylon is at approximately a 45-degree angle to your aircraft, roll out of the turn, and fly straight and level until the second pylon is directly off the wingtip; then roll into the second half of the maneuver. Flying time between the pylons, straight and level, should be about three to five seconds so you have time to clear the area of any other traffic. Remember, as you practice your pylon 8s, you will be very busy with your wingtip reference, so make very sure you use the time to clear the area wisely (FIG. 5-7).

The most common error pilots make in practicing this maneuver is to use their rudder to yaw the aircraft to hold the pylon on the wingtip reference point. While this works, it completely ruins the point of the maneuver and only results in an uncoordinated trip around the pylons. I teach my students to use their rudders only to roll into and out of the pylon 8. The rest of the time, I have them place their feet on the floor. This way, the bad habit of skidding the aircraft during the maneuver is never started.

I have heard student pilots talking to each other blasting the pylon 8 as a waste of time, money and fuel. And, to a point, I suppose they are correct. You will never pylon 8

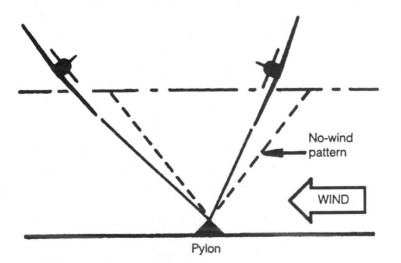

Fig. 5-5. *The wind always blows the aircraft downwind during a Pylon 8 because no correction is made for ground track.*

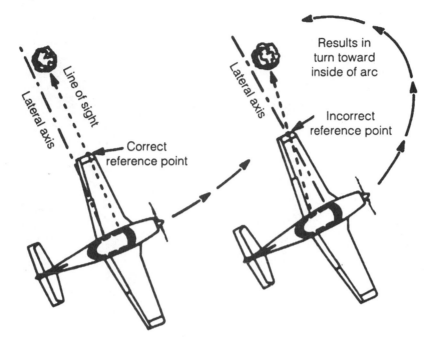

Fig. 5-6. *The rear seat occupant in a tandem seat aircraft has a tendency to turn toward the inside of the circle and descend because the view of the pylon will always make it look as if he is behind the reference line.*

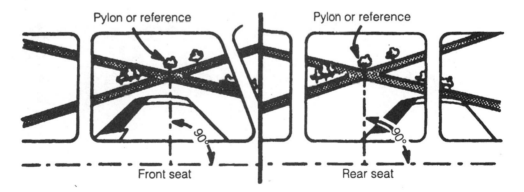

Fig. 5-7. *When using a tandem-type seating arrangement, the line of sight to the pylon is very different depending on whether you are in the front or rear seat.*

your way to Chicago or do a pylon 8 in the traffic pattern. But, the point they miss is the value of the pylon 8 in learning to fly the airplane by feel. The ability to safely aviate while one's attention is drawn outside is invaluable. And the coordination learned from varying the bank and pitch while maintaining a circular flight path is something that wouldn't hurt any pilot.

6
Cross-Country Flights

ALMOST ALL THAT YOU HAVE PREVIOUSLY LEARNED, PLUS A FEW new tips, will be put to use on a cross-country flight. To realize the full benefit from your flight training, you must be able to plan and execute flights to other areas. To safely accomplish a cross-country flight, you have to be familiar with your aircraft, sectional charts, trip planning, weather, and navigation in order to be able to navigate from one point to another.

A safe cross-country flight consists of two distinct phases: preflight planning and the flight itself. If your preplanning is thorough, you will be able to fly your trip with greater confidence and probably enjoy it a lot more. The more planning you do on the ground, the less busy you will be in the air. On your first few cross-country flights, you will have plenty of things to do to occupy your time and mind without trying to find that frequency, heading, or some other important detail that you should have already written down and put in a safe place.

PREFLIGHT PLANNING

A really safe cross-country flight doesn't just happen; it is planned. I remember my first dual cross-country flight. My instructor called me and said he had to go to Macon, Missouri, and would I like to go along. Would I? We got into the aircraft, flew to Macon, and returned. It was over before I knew it. He said I had done just fine, and he endorsed my certificate for solo cross-country privileges. There was just one problem: I did not know where we were going, where I had been, or how we had gotten there and back. There was virtually no preflight briefing, no calls to the FSS, no line on a chart (which he held the entire time), and only a general direction in which I was told to fly. At that time, I thought wind was something that kept you cool in the summer and a course was something that you took in school.

The odd part of this little self-incrimination was that the man was a very good basic flight instructor. He just didn't place much stock in the little things such as headings, charts, and frequencies. But he had taught me to fly the basics so well that I had time to look around, do some figuring, and with some common sense, I made it. The kicker is that I believe this made me a better pilot in many respects because I was forced to learn or suffer the consequences. I'm not recommending this as a way to learn cross-country flight technique, only suggesting that it can be done. If you receive good instruction, cross-country can be one of the more pleasant portions of your flight training.

The first things you need to know before you go on your cross-country are where you are going, the time of day you will be flying, and the charts that will be necessary to safely complete the trip (FIG. 6-1). You also have to plan whether the trip will be made by VOR, pilotage, or a combination of the two. Most VFR flights are made by using a combination of VOR and pilotage. However, with the advent of the FAA's changes to the FARs of 1997, pilotage is the method you must use to plan and execute your cross-country flight for your private pilot checkride.

Fig. 6-1. *Check the course and distance as part of your initial preparation.*

Pilotage

Pilotage means to navigate an aircraft by using a map and prominent landmarks on the surface of the earth. It is obviously the oldest form of navigation, dating back to the early 1900s. It is also one of the most certain methods for finding and maintaining your position relative to earthly landmarks since they appear around and below you in plain sight.

Today's pilot can utilize U.S. government-supplied sectional charts that depict the landmarks in very minute detail. In fact, there is so much information on these sectional charts that in a heavily populated area there is sometimes almost too much information. For instance, these charts display all towers, tell you how high they are above both sea level and the ground, as well as almost any other identifiable object you might imagine. A short list of items displayed on these charts would contain drive-in movie theatres, rivers, lakes, towns, towers, roads, airports, hills, mountains, cliffs, valleys, power plants, race tracks, railroad tracks, and just about any other solid object you can imagine. If you could see it, hit it, or use it, it will be displayed on the government sectional charts.

I have a small but significant piece of advice for you new or would-be pilots concerning the proper method of using the sectional chart. As is the case with all maps I am aware of, sectional charts are oriented to north. That is the top of the map, when held so it is easily readable, is north. Okay, so that's not earthshaking of itself. But human nature is such that we all like to be able to read things easily and so it is with sectional charts. We like to hold the map so that the map is always right side up. It's just easier to read that way.

Trouble is, our flights aren't necessarily always going from south to north. Sometimes we are going to fly southwest or whatever. When this happens, if you continue to hold the map so that it's right side up, the objects appear out the window of the aircraft out of place. They aren't in their proper perspective. Say a lake shows itself to your right, or west, of your course on the sectional. If you are flying south while holding the map upright, it's easy to believe the lake should appear out the left window of your aircraft. Of course this is not the case, but when you hold the map in what amounts to be an upside down position, funny things can happen to your navigation. Ask Wrong-Way Corrigan.

The cure for this malady is to always hold the map with the course you are following pointed out the front windshield of your aircraft. Turn the map so the departure point is nearest your body and the destination is at the point farthest away from you. It's only when landmarks appear out the window in the same relative positions in which they appear on the charts that the potential problem of reverse orientation is most likely avoided.

VOR

Without getting too complicated, let's review the functions and normal use of the very high frequency omnidirectional radio range (VOR). The VOR is called the *omni*. *Omni* is Latin for "in all directions." It can be a very good friend or a frightening enemy, depending on your understanding of its use and purpose. Well understood, it is

of invaluable assistance in navigation. If you try to fly the VOR without fully understanding its use, you might actually wind up going in the wrong direction.

From its ground base, the VOR transmits signals in all directions. Each VOR has its
own frequency, and you tune it in on the navigation side of your radio. The following describe the features of your VOR receiver.

Frequency selector

The *frequency selector* is manually rotated to select any of the frequencies in the VOR
range of 108.0 to 117.95 MHz.

Course selector

By turning the OBS (omni bearing selector), the desired course is selected. This usually
appears in a window or under an index on the VOR receiver head on your instrument
panel.

Course deviation indicator

(CDI) The *course deviation indicator* is composed of a dial and a needle. The needle
centers when you are on the selected course or its reciprocal, regardless of your
heading (FIG. 6-2). Full needle deflection from the center position to either side of
the dial indicates the aircraft is 10 degrees or more off course (assuming normal needle sensitivity).

TO/FROM Indicator

The *TO/FROM indicator* is also called *sense indicator* or *ambiguity indicator*. The
TO/FROM indicator shows whether the selected *course* will take the aircraft to or from the
station. It does not indicate whether the aircraft is heading to or from the station (FIG. 6-3).

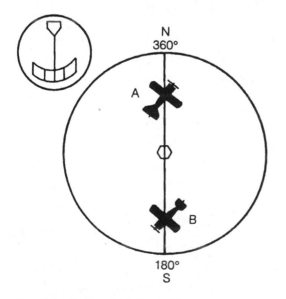

Fig. 6-2. *Course deviation indicator.*

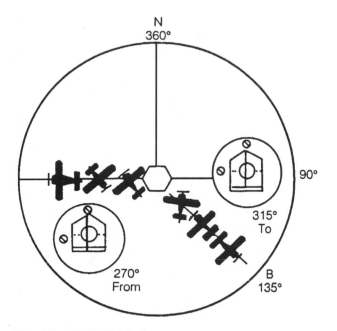

Fig. 6-3. *TO/FROM indicator.*

Flags

Flags can be labeled as signal strength indicators. The device to indicate whether a signal is usable or an unreliable signal is called an OFF/VOR flag. This flag retracts from view or says OFF when signal strength is sufficient for reliable instrument indications. When the VOR signal is strong enough to give reliable navigation guidance, the flag switches to VOR. Insufficient signal strength might also be indicated by a blank TO/FROM window.

There are a couple of very important points to remember when using the VOR. The VOR transmits 360 possible magnetic courses to and from the station. These courses are called *radials*. They are oriented *from* the station (FIG. 6-4). For example, the aircraft at A, heading 180 degrees, is flying to the station on the 360 radial. After crossing the station, the aircraft is flying from the station on the 180 radial at 2A. Aircraft B is shown crossing the 225 radial. Similarly, at any point around the station, an aircraft can be located somewhere on a VOR radial. The important point is if you want to know what radial you are on, turn the OBS until the CDI centers and the TO/FROM indicator reads FROM. That is the radial you are on.

To properly utilize your VOR, you must first know where you are (what radial you are on), where you are going (what radial you have to intercept), and how to track the radial once you get there. This process utilizes three steps: orientation, interception of the radial, and tracking.

The following demonstration can be used to intercept a predetermined inbound or outbound track. The first three steps may be omitted if you turn directly to intercept course without initially turning to parallel the desired course.

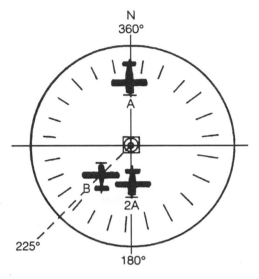

Fig. 6-4. *VOR courses called radials are oriented FROM the station.*

Turn to a heading to parallel the desired course. Turn in the same direction as the course to be flown.

- Determine the difference between the radial to be intercepted and the radial on which you are located.
- Double the difference to determine the intercept angle not less than 20 degrees or more than 60 degrees.
- Rotate the OBS to the desired radial or inbound course.
- Turn to the interception beading (magnetic).
- Hold this magnetic heading until the CDI centers, indicating the aircraft is crossing the desired course.
- Turn to the magnetic heading corresponding to the selected course and track inbound or outbound on the radial.

VOR tracking also involves drift correction sufficient to maintain a direct course to or from a station. The course selected for tracking inbound is the course shown on the course index with the TO/FROM indicator showing TO. If you are off course to the left, the CDI is deflected right; if you are off course to the right, the CDI is deflected to the left. Turning toward the needle returns the aircraft to the course centerline and centers the needle.

To track inbound with the wind unknown, proceed using the following steps (FIG. 6-5). Outbound tracking is the same.

- With the CDI centered, maintain the heading corresponding to the selected course.

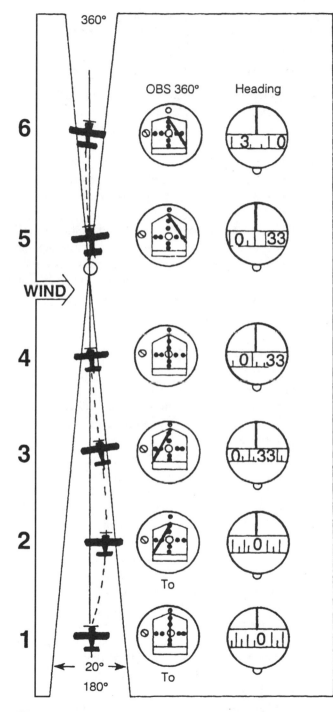

Fig. 6-5. *Tracking inbound with the wind unknown.*

- As you hold the heading, note the CDI for deflection to the left or right. The direction of CDI deflection from centerline shows you the direction of the crosswind. Figure 6-5 shows a left deflection, therefore a left crosswind.
- Turn 20 degrees toward the needle and hold the heading until the needle centers.

Reduce drift correction to 10 degrees left of the course setting. Note whether this drift-correction angle keeps the CDI centered. Subsequent left or right needle deflection indicates an excessive or insufficient drift-correction angle—either add or remove some correction (heading). With the proper drift correction established, the CDI will remain centered until the aircraft is close to the station. Approach to the station is shown by a flickering of the TO/FROM indicator and CDI as the aircraft flies into the no-signal area (almost directly over the station). Station passage is indicated by a complete reversal of the TO/FROM indicator.

Following station passage and TO/FROM reversal, course correction is still toward the needle to maintain course centerline. The only difference is that now you are tracking *away from* the station instead of *to* the station. In the previously listed steps, you were tracking inbound on the 180 radial. After station passage, although your heading hasn't changed, you are tracking outbound on the 360 radial.

Loran

Loran is an acronym for *LO*ng *RA*nge *N*avigation. An old form of navigation, Loran has gained unqualified acceptance with pilots. The advent of better and smaller computers has made the units more size and price competitive. Also, the FAA, until quite recently, was constructing more transmitting stations within the inner United States. Originally, Loran was utilized during World War II to assist with ocean navigation. Later, the Coast Guard further developed its usefulness with the newer Loran-C, which we now call Loran.

The operating principle of Loran is the measurement of time between two electronic impulses received from a chain of transmitters. The transmitters are spaced many hundreds of miles apart and are divided into master and secondary stations. Each station, master and secondary, transmits an impulse at precisely the same instant. The Loran receiver measures the difference in time that it takes for these impulses to travel to the receiver and computes its position based on these slight time differences. These position cross-checks are accurate to within one-quarter of a mile or better, and you'd better believe a position check from several hundred miles away that is that accurate is a welcome companion to any pilot. (FIG. 6-6.)

As computer technology advanced and the weight and size of receivers reduced, Loran receivers became more readily accessible to aircraft. The newer models are approximately the size of any other communication receiver. This sizing, along with more affordable pricing, has made the Loran a viable option to many pilots.

The Loran's small computer gives pilots an enormous amount of navigation information, so much so that some are intimidated by the Loran. Witness the information, ready at the pilot's fingertips, from a modern Loran:

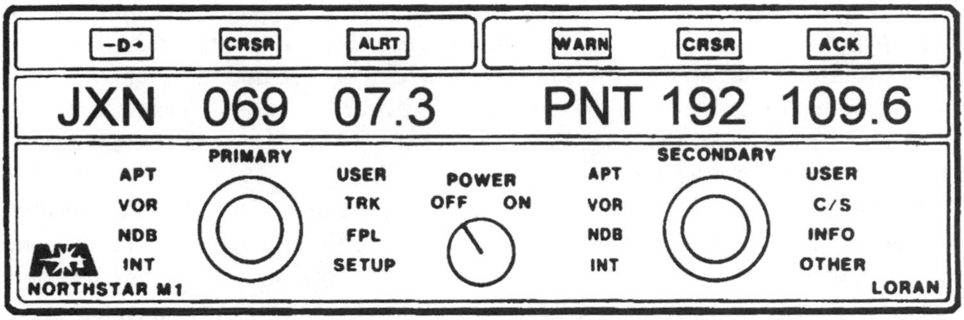

Fig. 6-6. *The LORAN can give a pilot a great deal of information.*

- Present position.
- Bearing and range to the station.
- Bearing and distance from current position to origination point.
- Estimated time enroute.
- Groundspeed.
- Distance from desired track and physical depiction of correction needed.

Additionally, the Loran can store in its memory thousands of bits of information concerning U.S. navigation aids, altitude restrictions, wind correction information, airports of the world, and much more. Is it any wonder this form of navigation has caught on with pilots in all walks of aviation?

The Loran does suffer a couple of points of detraction. It is very sensitive to precipitation, and proper antenna installation is critical to its accuracy. Also, the FAA is talking about reducing funding for Loran satellites and going with GPS (Global Positioning Systems)in the near future. But, personally, I doubt this will happen any time soon. The popularity of the Loran and the multiple tens of thousands of them in daily use make me believe the Loran will be with us for quite a few years. Should the FAA make any serious attempt to withdraw these inexpensive and user friendly little navigation tools from use, the outcry from pilots would be resounding. However, the FAA is giving quite a bit of priority to the GPS since it is much more accurate and has more possibilities for use than does the Loran.

The Loran chain

The 27 U.S. Loran transmitters that provide signal coverage for the continental U.S. and the southern half of Alaska are distributed from Caribou, ME, to Attu Island in the Aleutians. Station operations are organized into subgroups of four to six stations called "chains." One station in the chain is designated the "Master" and the others are "secondary" stations.

The Loran Navigation signal is a carefully structured sequence of brief radio frequency pulses centered at 100 kilohertz. The sequence of signal transmissions consists of a pulse group from the Master (M) station followed at precise time intervals by groups from the secondary stations, which are designated by the U.S. Coast Guard with the letters V. W. X. Y. and Z. All secondary stations radiate pulses in groups of eight, but the Master signal for identification has as additional ninth pulse.

The time interval between the recurrence of the Master pulse group is the Group Repetition Interval (GRI). The GRI is the same for all stations in a chain and each Loran chain has a unique GRI. Since all stations in a particular chain operate on the same radio frequency, the GRI is the key by which a Loran receiver can identify and isolate signal groups from a specific chain.

The line between the Master and each secondary station is the "baseline" for a pair of stations. Typical baselines are from 600 to 1,000 nautical miles in length. The continuation of the baseline in either direction is a "baseline extension."

Loran transmitter stations have time and control equipment, a transmitter, auxiliary power equipment, a building about 100 by 30 feet in size and an antenna that is about 700 feet tall. A station generally requires approximately 100 or more acres of land to accommodate guy lines that keep the antenna in position. Each Loran station transmits from 400 to 1,600 kilowatts of signal power.

The USCG operates 27 stations, comprising eight chains, in the United States. Four control stations, which monitor chain performance, have personnel on duty full time. The Canadian east and west coast chains also provide signal coverage over small areas of the NAS.

When a control station detects a signal problem that could affect navigation accuracy, an alert signal called "Blink" is activated. Blink is a distinctive change in the group of eight pulses that can be recognized automatically by a receiver so the user is notified instantly that the Loran system should not be used for navigation. In addition, other problems can cause signal transmissions from a station to be halted.

The Loran Receiver

Before a Loran receiver can provide navigation information for a pilot, it must successfully receive, or "acquire," signals form three or more stations in a chain. Acquisition involves the time synchronization of the receiver with the chain GRI, identification of the Master station signals from among those checked, identification of secondary station signals, and the proper selection of the point in each signal at which measurements should be made.

Signal reception at any site will require a pilot to provide location information such as approximate latitude and longitude, or the GRI to be used, to the receiver. Once activated, most receivers will store present location information for later use. The basic measurements made by Loran receivers are the difference in time-of-arrival between the Master signal and the signals from each of the secondary stations of a chain.

An aircraft's Loran receiver must recognize three signal conditions:

- Usable signals.
- Absence of signals.
- Signal blink.

The most critical phase of flight is during the approach to landing at an airport. During the approach phase the receiver must detect a lost signal, or a signal Blink, within 10 seconds of the occurrence and warn the pilot of the event.

Loran signals operate in the low frequency band around (100 kHz) that has been reserved for Loran use. Adjacent to the band, however, are numerous low-frequency communications transmitters. Nearby signals that can distort the Loran receivers have selective internal filters. These filters, commonly known as "notch filters" reduce the effect of interfering signals.

Careful installation of antennas, good metal-to-metal electrical bonding, and provisions for precipitation noise discharge on the aircraft are essential for the successful operation of Loran receivers. A Loran antenna should be installed in accordance with the manufacturer's instructions. Corroded bonding straps should not be used, and static discharge devices installed at points indicated should be by the aircraft manufacturer.

Loran Navigation

An airborne Loran receiver has four major parts:

- Signal processor.
- Navigation computer.
- Control/display.
- Antenna.

The signal processor acquires Loran signals and measures the difference between the time-of-arrival of each secondary station pulse group and the Master station pulse group. The measured TDs depend on the location of the receiver in relation to the three or more transmitters.

The first TD will locate an aircraft somewhere on a line-of-position (LOP) on which the receiver will measure the same TD (time-distance) value. A second LOP is defined by a TD measurement between the Master station signal and the signal from another secondary station. The intersection of the measured LOPs is the position of the aircraft.

The navigation computer converts TD values to corresponding latitude and longitude. Once the time and position of the aircraft is established at two points, distance to destination, cross track error, ground speed, estimated time of arrival, etc., can be determined. Cross-track error can be displayed as the vertical needle of a course deviation indicator, or digitally, as decimal parts of a mile left or right of course. During a nonprecision approach, course guidance must be displayed to the pilot with a full-scale deviation of 0.30 nautical miles or greater.

Loran navigation for nonprecision approaches requires accurate and reliable information. During an approach the occurrence of signal Blink or loss of signal must be detected within 10 seconds and the pilot must be notified. Loran signal accuracy for approaches is 0.25 nautical miles, well within the required accuracy of 0.30 nautical miles.

Flying a Loran nonprecision approach is different from flying a VOR approach. A VOR approach is on a radial off the VOR station, with guidance sensitivity increasing as the aircraft nears the airport. The Loran system provides a linear grid, so there is constant guidance sensitivity everywhere in the approach procedure. Consequently, inaccuracies and ambiguities that occur during operations in close proximity to VORs (station passage, for example) do not occur in Loran approaches.

The navigation computer also provides storage for data entered by pilot or provided by the receiver manufacturer. The receiver's database is updated at local maintenance facilities every 60 days to include all changes made by the FAA.

GLOBAL POSITIONING SYSTEM (GPS)

GPS is a U.S. satellite-based radio navigational, positioning and time-transfer system operated by the Department of Defense (DOD). The system provides highly accurate position, velocity, and precise time information on a continuous global basis to an unlimited number of properly equipped users. Unlike LORAN, the GPS system is unaffected by weather and provides a worldwide common grid reference system based on an earth-fixed coordinate system.

GPS provides two levels of service: Standard Positioning Service (SPS) and Precise Positioning Service (PPS). SPS provides, to all users, horizontal positioning accuracy of 10 meters, or less, with a probability of 95 percent and 300 meters with probability of 99.99 percent. PPS is more accurate than SPS: However, this is limited to authorized U.S. civil and military users who meet specific requirements.

GPS operation is based on the concept of ranging and triangulation from a group of satellites in space which act as precise reference points. A GPS receiver measures distance from a satellite using the travel time of a radio signal. Each satellite transmits a specific code, called a course/acquisition (CA) code, which contains information on the satellite's position, the GPS system time, and the health and accuracy of the transmitted data. Knowing the speed at which the signal traveled, the speed of light, (approximately 186,000 miles per second), and the exact broadcast time, the distance traveled by the signal can be computed from the arrival time.

The GPS receiver matches each satellite's CA code with an identical copy of the code contained in the receiver's database. By shifting its copy of the satellite's code in a matching process, and by comparing this shift with its internal clock, the receiver can calculate how long it took the signal to travel from the satellite to the receiver. The distance derived from this method of computing distance is called a pseudo-range because it is not a direct measurement of distance, but a measurement based on time.

In addition to knowing the distance to a satellite, a receiver needs to know the satellite's exact position in space; this is known as its ephemeris. Each satellite transmits information about its exact orbital location. The GPS receiver uses this information to precisely establish the position of the satellite.

Using the calculated pseudo-range and position information supplied by the satellite, the GPS receiver mathematically determines its position by triangulation. The GPS receiver needs at least four satellites to yield a three-dimension position (latitude, longitude, and altitude) and time solution. The GPS receiver computes navigational values such as distance and bearing to a waypoint, groundspeed, etc., by using the aircraft's known latitude/longitude and referencing these to data built into the receiver.

The GPS constellation of 24 satellites is designed so that a minimum of five is always observable by a user anywhere on earth. The receiver uses data from a minimum of four satellites above the mask angle (the lowest angle above the horizon at which it can use a satellite).

The GPS receiver verifies the integrity of the signals received from the GPS constellation through receiver autonomous integrity monitoring (RAIM) to determine if a satellite is providing corrupted information. At least one satellite, in addition to those required for navigation, must be in view for the receiver to perform the RAIM function; thus, RAIM needs a minimum of 5 satellites in view, or 4 satellites and a barometric altimeter (baro-aiding) to detect an integrity anomaly. For receivers capable of doing so, RAIM needs 6 satellites in view (or 5 satellites with baro-aiding) to isolate the corrupt satellite signal and remove it from the navigation solution. Baro-aiding is a method of augmenting the GPS integrity solution by using a nonsatellite input source. GPS derived altitude should not be relied upon to determine aircraft altitude since the vertical error can be quite large. To ensure that baro-aiding is available, the current altimeter setting must be entered into the receiver as described in the operation manual.

The Department of Defense declared initial operational capability of the U.S. GPS on December 8, 1993. The Federal Aviation Administration (FAA) has granted approval for U.S. civil operators to use properly certified GPS equipment as a primary means of navigation in oceanic airspace and certain remote areas. Properly certified GPS equipment may be used as supplemental means of IFR navigation for domestic enroute, terminal operations, and certain instrument approach procedures. This approval permits the use of GPS in a manner that is consistent with current navigation requirements as well as approved air carrier operations specification.

Civilian pilots are authorized to conduct any GPS operation under IFR provided that the GPS navigation equipment used is approved in accordance with the requirements specified in TSO C-129, or equivalent, and the installation must be done in accordance with Notice 8110.47 or 8110.48, or the equivalent.

Procedures must be established for use in the event that the loss of RAIM capability is predicted to occur. In situations where this is encountered, the flight must rely on other approved navigation equipment (VOR, ADF, etc.), delay departure, or cancel the flight. (FIG. 6-7.)

The GPS operation must be conducted in accordance with the FAA-approved aircraft flight manual or flight manual supplement. Flight crewmembers must be thoroughly familiar with the particular GPS equipment installed in the aircraft, the receiver operation manual, and the AFM or flight manual supplement. Unlike ILS and VOR, the basic operation, receiver presentation to the pilot, and some capabilities of

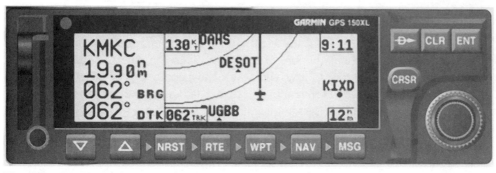

Fig. 6-7. *The GPS is the FAA's chosen method of navigation now and into the 21st century.*

the equipment can vary greatly. Due to these differences, operation of different brands, or even models of the same brand, of GPS receiver under IFR should not be attempted without thorough study of the operation of that particular receiver and installation. Most receivers have a built-in simulator mode which will allow the pilot to become familiar with operation prior to attempting operation in the aircraft. I would suggest using the equipment in flight under VFR conditions prior to attempting IFR operation.

Aircraft navigation by IFR approved GPS are considered to be RNAV aircraft and have special equipment suffixes as related to ATC flight plans. You must file the appropriate equipment suffix on the ATC flight plan. If GPS avionics become inoperative, the pilot should advise ATC and amend the equipment suffix.

GPS position orientation

As with most RNAV systems, pilots should pay close attention to position orientation while using GPS. Distance and track information are provided to the next active waypoint, not to a fixed navigation aid. Receivers may sequence when a pilot is not flying along an active route, such as when being vectored or deviating due to weather, due to the proximity to another waypoint in the route. This can be prevented by placing the receiver in a nonsequencing mode. When the receiver is in the nonsequencing mode, bearing and distance are provided to the selected waypoint and the receiver will not sequence to the next waypoint in the route until placed back in the auto sequence mode or the pilot selects a different waypoint. On an overlay approach, the pilot may have to compute the track distance to step-down fixes and other points due to the receiver showing track distance to the next waypoint rather than DME to the VOR or ILS ground station.

GPS versus conventional navigation

You may find slight differences between the heading information shown on the navigational charts and the GPS navigation display when flying on an overlay approach or along an airway. All magnetic tracks defined by a VOR radial are determined by the application of magnetic variation at the VOR. However, GPS operations may use an algorithm to apply the magnetic variation at the current position, which may produce small differences in the displayed course. Both operations should produce the same desired ground track. Due to the use of great circle courses, and the variations in magnetic variations, the bearing to the next waypoint and the course from the last waypoint may not be exactly 180 degrees apart when long distances are involved.

Variations in distances will occur since GPS distance-to-waypoint values are along track (straight-line) distances computed to the next waypoint and the DME values published on underlying procedures are the slant range distances measured to the station. The difference increases with aircraft altitude and proximity to the NAVAID.

Waypoints

GPS approaches make use of both flyover and fly-by waypoints. Fly-by waypoints are used when an aircraft should begin a turn to the next course prior to reaching the waypoint separating the two route segments. This is known as turn anticipation and is compensated for in the airspace and terrain clearances. Approach waypoints, except for the missed approach waypoint (MAWP), and the missed approach holding waypoint (MAHWP), are normally fly-by waypoints. Flyover waypoints are used when the aircraft must fly over the point prior to starting a turn. New approach charts depict flyover waypoints as a circled waypoint symbol. Overlay approach charts and some early stand-alone GPS approach charts may not reflect this convention.

On overlay approaches (titled "or GPS"), if no pronounceable five character name is published for an approach waypoint or fix, it may be given an ARINC database identifier consisting of letters and numbers. These points will appear in the list of waypoints in the approach procedure database, but may not appear on the approach chart. Procedures without a final approach fix (FAF), for instance, have a sensor final approach waypoint (FAWP) added to the database at least 4 NM prior to the MAWP and MAHWP to allow the receiver to transition to the approach mode. Some approaches also contain an additional waypoint in the holding pattern when the MAWP and MAHWP are co-located. Arc and radial approaches have an additional waypoint that is used for turn anticipation computation when the arc joins the final approach course. These coded names will not be used by ATC.

Unnamed waypoints in the database will be uniquely identified for each airport but may be repeated for another airport (e.g., RW36 will be used at each airport with a runway 36 but will be at the same location for all approaches at a given airport).

The runway threshold waypoint, which is normally the MAWP, may have a five letter identifier (e.g., SNEEZ) or be coded as RW## (e.g., RW36, RW36L). Those thresholds which are coded as five letter identifiers are being changed to the RW## designation. This may cause the approach chart and database to differ until all changes are complete. The runway threshold waypoint is also used as the center of the MSA on most GPS approaches. MAWPs not located at the threshold will have a five-letter identifier.

Chapter Six

CHARTING THE COURSE

In order to utilize your VOR, Loran, GPS, or pilotage technique on cross-country flight, you need a chart to plot your trip. The sectional aeronautical charts are the ones used most often by pilots flying a VFR flight. These charts divide the country into 37 different parts, each named for a major city that lies within the boundary of each chart (FIG. 6-8).

However, the first thing you should do is go to the large wall chart at your airport and locate your point of departure. With the string attached to the compass rose in the center, locate your destination and obtain your general course and distance. Then, select the sectional chart that will suit your planned flight.

Let's say you are going from Lawrenceville, Illinois, to Evansville, Indiana. For this trip, the St. Louis Sectional fits the bill, as the trip is well within the boundaries of this particular chart.

Open the St. Louis Sectional and locate the two airports. Using a straightedge, draw a line from the center of the departure airport to the center of the destination airport. You will want to circle prominent checkpoints along your route of flight. Select ones that can be easily seen and recognized from the air. Then, use your plotter and obtain the true course for the cross-country. In this case, the true course is 175 degrees. You will also want to use your personal computer to access weather information or call the FSS to find

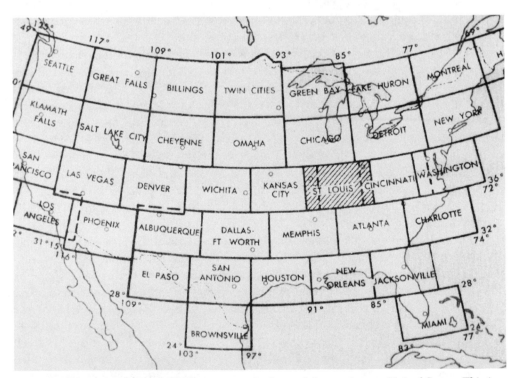

Fig. 6-8. *Thirty-seven sectional charts cover the forty-eight contiguous United States. This image is depicted on the front of each chart and helps in selecting the chart you need.*

out what type of weather to expect along your route as well as obtain the forecast winds aloft.

Using your flight computer, find your true airspeed, true heading, groundspeed, and magnetic and compass headings. The easy way to remember the sequence for this computation is the initials TVMDC. *TVMDC* stands for true, variation, magnetic, deviation, and compass. These initials are sometimes remembered in pilot circles as "True Virgins Make Dull Company."

Use both the St. Louis Sectional and the Aeronautical Information Manual as well to obtain radio frequencies, airport data, NOTAMs, new and closed airports, and any other information relevant to your proposed flight. Be sure to write this information down so you can find it quickly. Many pilots use a flight log that can be found on the back of FAA Flight Plans. Some merely use a piece of paper. In any case, don't try to commit it to memory-write it down.

When you have gathered all the pertinent data, fill out and file a VFR flight plan with the nearest FSS. As I said at the beginning of this book, this filing is the cheapest insurance you can have, so don't neglect it.

As you taxi out, set your OBS to 175 degrees and tune in the Lawrenceville VOR on 108.8. Arrange all your flight plans and notes and fold your Sectional Chart so only the portion you are using shows. As I said before, turn your Sectional Chart so that it corresponds with your route of flight. Place it on your lap with your destination airport furthest away from your body. When you look outside, everything will then be in the proper perspective. A town that should be off to your right will be off to your right. You will learn to read upside down very quickly. Keep the sectional turned in the direction of your course.

I have advised you to set your VOR and keep your sectional in place at the same time. Then, you can fly the VOR and double-check your route with the sectional, or vice versa. A strong word of caution: If you ever decide to fly either VOR or *pilotage* exclusively, fly pilotage (the map). Radios have a very peculiar habit of malfunctioning at the most inopportune times. Know where you are on the sectional. I've known a couple of pilots who liked to fly the VOR with the sectional somewhere in the back seat, out of sight, and then actually flew right over their intended destination and never even knew it. I guess it's all right if you're trying to build flight time because you will certainly have to turn around and hunt for your destination.

THE FLIGHT

Once airborne, make sure you get off to a good start on your cross-country by being on course from the very beginning. In order to assure this, some pilots take off, circle the airport, and fly directly over the departure airport so they are absolutely sure they are right on target when they part. Not a bad idea, at first.

Headings

If your plan calls for flying both the VOR and pilotage, you should take up a heading that allows you to intercept the 175-degree radial from the Lawrenceville VOR. At the same time, you should begin to pick up your visual checkpoints. In this case, the first good one is when you cross the four-lane highway just south of the Lawrenceville airport. From this position, the city of Vincennes should be visible off to your left and slightly ahead. The city of Lawrenceville will be just about under your right wingtip. You are off to a good start, and it is only a matter of keeping close track of the other checkpoints as they come up. You can crosscheck your position by reference to the CDI indication. If you are directly on course, it will read FROM the 175-degree radial of the Lawrenceville VOR, and the needle will be in the center.

Visual checkpoints

As you continue on towards Evansville (FIG. 6-9), the next set of visual checkpoints will be about 5 miles south of the four-lane highway, where you should see the bridge that crosses the Wabash River at St. Francisville about a mile to the right of your course. On a fairly clear day, you should be able to see the Mt. Carmel airport a little further to the west. You're still on course. It's a good feeling when the checkpoints come up at the proper time and in the proper place.

If you want to do a crosscheck with the VOR to really confirm your exact position, tune in the Evansville VOR shown in the bottom left corner of the picture of the St. Louis Sectional in FIG. 6-9 on the frequency shown in the box just under the northern most edge of the VOR compass rose. This frequency is 113.3. Let's say you are crossing over the city of Princeton. With the frequency set in, rotate the OBS until the needle centers on a FROM indication. (Remember, you find the radial you are on with a FROM indication.) It should read 017 degrees FROM as you pass over Princeton. Now, you know exactly where you are. This method can be utilized anywhere, at any time, as long as you are receiving a strong enough signal to know the VOR is reliable. You know you are receiving a usable signal when the TO/FROM indicator is firmly on either TO or FROM, not bouncing from one to the other, and certainly never when the flag box reads OFF.

ATIS frequency

Before you left, you checked the Aeronautical Information Manual and found the frequencies you would need for the flight. One of them should have been the ATIS (Automatic Terminal Information Service) frequency for Evansville. If you forgot to obtain this information, it can be found in the Airport Traffic Area information listed just to the northeast of the Evansville airport. The Evansville ATIS broadcasts on 120.2 on the communications side of your radio. This transcribed information gives you a great deal of information relating to weather, runways in use, frequencies to contact from different arrival positions, and more. It broadcasts continuously, is updated every hour, and is coded such as, "Evansville, Dress Regional Airport, information Bravo." The last thing the ATIS will usually tell you is, "Inform controller you have information Bravo." You should begin to listen to the ATIS at least 25 miles out. Therefore, over Princeton would be a good time to tune in to see what is happening at Evansville.

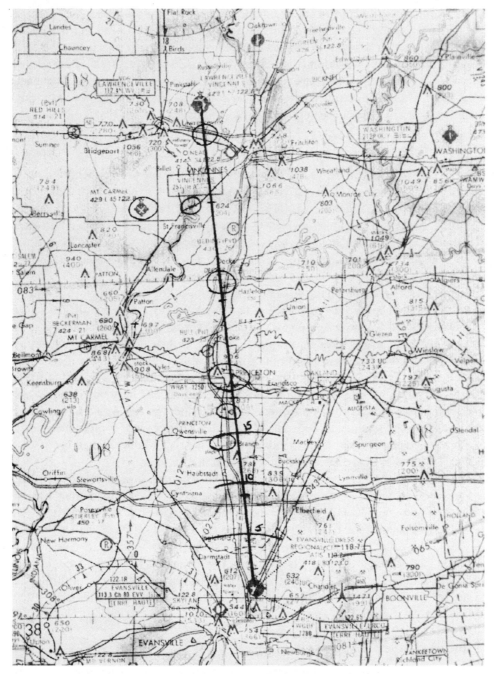

Fig. 6-9. *This sample cross-country, shown on the St. Louis Sectional, is from Lawrenceville, Illinois to Evansville, Indiana.*

After you have written down any pertinent information you have received from the ATIS, call the approach control and tell them who you are, where you are, and what your intentions are. In other words, you might say, "Evansville approach, this is Cessna 152, N94469, 15 miles north, with information Bravo, landing Evansville. Over." If the airport is equipped with radar, they will have you tune your transponder to a certain frequency for radar identification. If you are not transponder-equipped, tell them and they will probably have you turn to a given heading for radar identification.

The approach

The approach controller will tell you to plan on making a straight-in approach, right traffic, or whatever for the runway in use. Approach will tell you to contact the tower at a certain point, often three miles out. If you have time, write down what he tells you to do. It saves time, but if you can't manage to write and fly at the same time, fly and do your best to remember the important things like which runway you are to use and the type of approach you will make. It is better to keep flying and have to call the controller again for information than to try to write and lose control of your aircraft.

When you arrive at the point the approach control told you to contact the tower, switch over to the tower frequency (118.7 in this case) and give them a call. The tower controller most likely knows where you are and will give you instructions accordingly. It might be a simple, "Cleared to land, runway 18." Whatever they tell you, be sure that you understand. If you don't understand, ask the controller to say it again. Safety must come first. I have many thousands of flight hours, yet I still sometimes have trouble understanding what has been said or the intent they have in mind. If I have any doubt, I ask again. Don't be shy. Remember, controllers are people too, just like you and me. They make mistakes too. If you feel the clearance you have been given will compromise a safe flight, do what you have to do and worry about "the voice" later. Remember, it is not God speaking. It is another mere human, just like you and me.

I once had a controller clear me for a straight-in approach to a runway. About the same time he cleared a DC-9 for a straight-in for the same runway. I was on short final in a Cessna 310 when the DC-9 passed over me executing his go-around. He had not seen me until the last instant, and none of us, including me, the controller, or the DC-9 pilot was aware of the situation. When we had the inevitable meeting with the tower chief after all had landed safely, it came out that the controller thought he had cleared me for runway 35 right and the DC-9 for runway 35 left. The tapes showed otherwise. But the sad part was that three human beings all made a potentially dreadful error, and did it at the same time. This is exactly how accidents happen.

Pilot/controller cooperation

A great deal has been written about the sometimes-adversarial relationship between pilots and air traffic controllers. If you listen to one side, the controller is the mortal enemy of the pilot, that authoritative voice that emanates from the headset and attempts to steer you into another aircraft, mountain, or thunderstorm. And if you listen to the other side, a pilot is a person sitting in an aluminum tube waiting for ATC to guide him since he cannot find his way on his own. The truth is somewhere in between, that both are consummate professionals attempting to do a difficult job in a crowded sky. And most do it very well.

I have had my problems with ATC, and ATC has had its problems with me. As a rule, it winds up to be an error in communication. One of us has misinterpreted the intentions of the other. Granted, this can lead to some very delicate moments, but we are, after all, human. This is the thrust of the issue; we are all fallible and we need to keep this in mind as we fly. We need to trust each other—but only to a point. We must all rely on the fact that we all make mistakes and that the other guy is there to make us aware of these errors as well as to work with us. An old friend of mine says it best: "Life is best when your takeoffs and your landings add up to an even number." Think about it.

Some controllers can even be fun. Witness: When the weather is below VFR minimums during the daytime, the rotating beacon on the control tower is supposed to be turned on, signifying that the weather is IFR. Upon arriving for work one very bleak and dismal morning, every pilot in turn noted the beacon on the control tower, which was located right next to us, was not lit. For some reason, it was as if all was not right with our world. It bothered us and we kept asking each other why the beacon was not on. Was it burned out? Was it an oversight?

Finally, I couldn't stand it any more and phoned the tower. "What's the matter with you guys? Why isn't the beacon on?" I asked. "Is it burned out? You know we can't get our morning started until we see the beacon."

"Look again," came the reply over the phone. "You'll see a beacon, all right."

We moved to the windows, not prepared for the sight that greeted us. There at the top of the tower, surrounded by an acre of solid glass, a controller who had jumped up onto the ledge inside the tower and dropped his drawers gave us a full-blown, ass-to-the-glass, moonshot. It was hysterical.

Not to be outdone we each grabbed blank pieces of paper on which we wrote scores, went outside, lined up like they do in the Olympics, and held the scores up for the controller to see. What marvelous camaraderie: pilot and controller playing together, and a few minutes later, working together. The next time you begin to feel shy about talking to an air traffic controller, remember this story. It should help.

CONCLUDING THE CROSS-COUNTRY FLIGHT

Once on the ground at Evansville, turn off of the active runway as soon as it can be done safely, or continue to follow the directions from the tower if you are able to comply with their instructions safely.

When clear of the active runway and just past the triple yellow hold lines, stop and call the ground controller for your instructions for taxi. Be sure to tell him you are clear of the active runway and where you want to go. If the airport layout is unfamiliar to you, ask the ground controller for a progressive taxi. This means they will give you step-by-step directions to where you need to go. Getting you around a strange airport and to your desired destination is what they get paid to do and they're usually very glad to help (See FIG. 6-10).

Fig. 6-10. *A safe arrival after a well-executed cross-country is a feeling of real satisfaction.*

Fig. 6-11. *A flight plan may be closed by radio, telephone, or in person, but don't forget to close it!*

Upon reaching your destination, shut down your aircraft, gather your charts, plotter, etc., tie your aircraft down, go into the terminal, and *CLOSE YOUR FLIGHT PLAN*. Do not go to the restroom, get a drink, or anything else until you have closed your flight plan with the FSS. (FIG. 6-11.)

This short cross-country is a typical example. They are all basically the same. Whether you are going 50 or 500 miles, you must plan for it as best you can and then fly it as best you can, keeping safety in the foremost portion of your mind. Many pilots tend to get a little uptight during the cross-country portion of flying. When they do, accidents happen with much more regularity. Take your time, relax, and use your common sense if you want to get the most out of cross-country flying.

7
Federal Airspace System

THE *FEDERAL AIRSPACE SYSTEM* IS AN IMPOSING PROPOSITION TO many pilots, fledging and experienced alike. The altitude limits, transition areas, visibility requirements, flight levels, etc., can be overwhelming if you try to view it all at once. Actually, there really isn't anything to view. There are only invisible zones, unseen routes, and those voices that come from who-knows-where. It can be disconcerting until you learn to separate the fact from the hangar talk.

IN GENERAL

Airspace is divided into two basic groups: regulated and unregulated, or controlled and uncontrolled as most pilots view it. I believe that many novice pilots' anxiety emanates from the word *controlled*. It sounds so final, so authoritative. In my youth, it kept me away from a lot of places I really wanted to be until I learned what I was missing. But, like most other seemingly difficult endeavors, if you stand back and look at it a piece at a time, it is not so intimidating.

Within the two categories of airspace, there are four specific types of airspace:

- Controlled.
- Uncontrolled.

- Special use.
- Other.

What dictates how these categories and types of airspace are utilized is the complexity and density of aircraft movements, the nature of the operations conducted within the airspace, the level of safety required and the national and public interest. It is important that pilots be familiar with the operational requirements for each of the various types or classes of airspace.

CONTROLLED AIRSPACE

In 1994, the FAA changed all federal airspace to the new alphabet airspace system. Instead of control zones and airport traffic areas, we now have six classes of airspace, each with its own alphabet initial denoting an airspace with different procedures, services and rules. The six airspace classes are; A, B, C, D, E, and G. As a primer to airspace regulations, just remember that airspace classes A, B, C, D, and E are thought of as controlled airspace and only class G is thought of as uncontrolled airspace.

Within controlled airspace, IFR and VFR flights operate together, separated by Air Traffic Control (ATC), through various sets of rules, regulations, and just good habits. For instance, IFR operations in any class of controlled airspace requires that a pilot file an IFR flight plan and receive an appropriate ATC clearance prior to entering IFR weather. Standard IFR separation is provided to all aircraft operating under IFR in controlled airspace.

In VFR flight, it is the responsibility of the pilot to ensure that ATC clearances or radio communication requirements are met prior to entry into Class B, Class C, or Class D airspace. The pilot retains this responsibility even when receiving ATC radar advisories. This means the responsibility to see and avoid other aircraft remains with the VFR pilot-in-command at all times, even when receiving traffic advisories from ATC. (FIG. 7-1.)

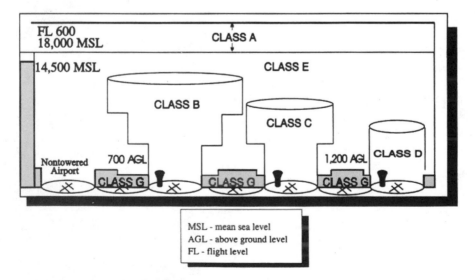

Fig. 7-1. *Airspace Classifications*

Fig. 7-2. *This Boeing 767 is the type of aircraft most commonly found in Class B Airspace.*

CLASS A AIRSPACE

Class A airspace is generally defined as that airspace from 18,000 feet MSL up to and including FL 600, including the airspace overlying the waters within 12 nautical miles of the coast of the 48 contiguous States and Alaska. Class A airspace also contains designated international airspace beyond 12 nautical miles of the coast of the 48 contiguous States and Alaska within areas of domestic radio navigational signal or ATC radar coverage, and within which domestic procedures are applied.

Class A airspace is up where the big boys fly. In Class A the aircraft have to be operating on an IFR flightplan and have an operating Mode C transponder. Class A airspace is kind of like your brother-in-law, it doesn't appear in any way on aviation charts, nor does it mean much to most of us, it's just there.

CLASS B AIRSPACE

Class B airspace is generally that airspace from the surface to 10,000 feet MSL surrounding the nation's busiest airports in terms of IFR operations or passenger enplanements. (FIG. 7-2.)

The configuration of each Class B airspace area is individually tailored and consists of a surface area and two or more layers (some Class B airspace areas resemble upside-down wedding cakes), and is designed to contain all published instrument procedures once an aircraft enters the airspace. Class B airspace is charted on Sectional Charts, IFR En Route Low Altitude, and Terminal Area Charts.

An ATC clearance is required for all aircraft to operate in the area, and all aircraft that are so cleared receive separation services within the airspace. The cloud clearance and in-flight visibility requirements for VFR operations within Class B airspace are 3 miles visibility and remain clear of clouds.

Class B has its own operating rules and pilot equipment requirements for VFR operations. Irregardless of weather conditions, an ATC clearance is required prior to operating within Class B airspace. Pilots should not request a clearance to operate within Class B airspace unless the requirements of FAR Part 91.215 and FAR Part 91.131 are met. These requirements include, unless otherwise authorized by ATC, the aircraft be equipped with an operable two-way radio capable of communicating with ATC on appropriate frequencies for that Class B airspace. Additionally, an operating mode C transponder is required prior to entering the airspace.

Because Class B contains the really busy airports in our country, the FAA desires that no student pilots be allowed to fly within the Class B airspace unless they have had dual instruction and are signed off for flight in a specific Class B airspace. So, no person may take off or land a civil aircraft at the primary airports within Class B airspace unless the pilot in command holds at least a private pilot certificate. This doesn't mean that students cannot fly under the Class B airspace and do anything they want, anywhere they want. They just have to remain clear of the Class B airspace unless they have the Class B endorsement.

One other facet peculiar to Class B is the primary airport is surrounded by an invisible 30-mile diameter ring called a Mode C veil. This veil extends from the surface up to 10,000 feet MSL and no one may fly within this veil without an operating Mode C transponder. Note: ATC may, upon notification, immediately authorize a deviation from the altitude reporting (Mode C) requirement. However, a request for a deviation from the 4096 transponder equipment requirement must be submitted to the controlling ATC facility at least one hour before the proposed operation.

There is but one exception to the Mode C veil rule. Older aircraft that were not originally certificated with an engine-driven electrical system or which have not subsequently been certified with a system installed may conduct operations within a Mode C veil, provided the aircraft remains outside Class A, B, or C airspace and below the altitude of the ceiling of a Class B or Class C airspace area designated for an airport or 10,000 feet MSL, whichever is lower.

There are 12 airports in the conterminous U.S. designated as Class B airports because of their traffic volume. They are:

- Andrews Air Force Base, MD
- Atlanta Hartsfield Airport, GA
- Boston Logan Airport, MA
- Chicago O'Hare Intl Airport, IL
- Los Angeles Intl Airport, CA

- Miami Intl Airport, FLA
- Newark Intl Airport, NJ
- New York Kennedy Airport, NY
- New York La Guardia Airport, NY
- San Francisco Intl Airport, CA
- Washington National Airport, DC
- Dallas/Fort Worth Intl Airport, TX

Flight procedures

All aircraft within Class B airspace are required to operate in accordance with current IFR procedures. A clearance for a visual approach to a primary airport is not authorization for turbine powered airplanes to operate below the designated floors of the Class B airspace.

VFR flights arriving aircraft must obtain an ATC clearance prior to entering Class B airspace and must contact ATC on the appropriate frequency, and in relation to geographical fixes shown on local charts. Although a pilot may be operating beneath the floor of the Class B airspace on initial contact, communications with ATC should be established in relation to the points indicated for spacing and sequencing purposes.

Departing aircraft require a clearance to depart Class B airspace and should advise the clearance delivery position of their intended altitude and route of flight. ATC will normally advise VFR aircraft when leaving the geographical limits of the Class B airspace. Radar service is not automatically terminated with this advisory unless specifically stated by the controller.

Aircraft not landing or departing the primary airport may obtain an ATC clearance to transit the Class B airspace when traffic conditions permit. Such VFR aircraft are encouraged, to the extent possible, to operate at altitudes above or below the Class B airspace or to transit through established VFR corridors. Pilots operating in VFR corridors are urged to use frequency 122.750 MHz for the exchange of aircraft position information.

An ATC clearance is required to enter and operate within Class B airspace. VFR pilots are provided sequencing and separation from other aircraft while operating within Class B airspace. NOTE: Separation and sequencing of VFR aircraft will be suspended in the event of a radar outage as this service is dependent on radar. The pilot will be advised that the service is not available and issued wind, runway information, and the time or place to contact the tower.

Radar separation should not be interpreted as relieving pilots of their responsibilities to see and avoid other traffic operating in basic VFR weather conditions (see FIG. 7-3), to adjust their operations and flight path as necessary to preclude serious wake encounters, to maintain appropriate terrain and obstruction clearance or to remain in weather conditions equal to or better than the VFR weather minimums.

Approach control should be advised and a revised clearance or instruction obtained when compliance with an assigned route, heading or altitude is likely to compromise pilot responsibility with respect to terrain and obstruction clearance, vortex exposure, and weather minimums.

Basic VFR Weather Minimums

Airspace	Flight Visibility	Distance from Clouds
Class A	Not Applicable	Not Applicable
Class B	3 statute miles	Clear of Clouds
Class C	3 statute miles	500 feet below 1,000 feet above 2,000 feet horizontal
Class D	3 statute miles	500 feet below 1,000 feet above 2,000 feet horizontal
Class E Less than 10,000 feet MSL	3 statute miles	500 feet below 1,000 feet above 2,000 feet horizontal
At or above 10,000 feet MSL	5 statute miles	1,000 feet below 1,000 feet above 1 statute mile horizontal
Class G 1,200 feet or less above the surface (regardless of MSL altitude).		
Day, except as provided in section 91.155(b)	1 statute mile	Clear of clouds
Night, except as provided in section 91.155(b)	3 statute miles	500 feet below 1,000 feet above 2,000 feet horizontal
More than 1,200 feet above the surface but less than 10,000 feet MSL.		
Day	1 statute mile	500 feet below 1,000 feet above 2,000 feet horizontal
Night	3 statute miles	500 feet below 1,000 feet above 2,000 feet horizontal
More than 1,200 feet above the surface and at or above 10,000 feet MSL.	5 statute miles	1,000 feet below 1,000 feet above 1 statute mile horizontal

Fig. 7-3. *Basic VFR weather minimums.*

ATC may assign altitudes to VFR aircraft that do not conform to FAR Part 91.159 concerning VFR flight altitudes. "RESUME APPROPRIATE VFR ALTITUDES" will be broadcast when the altitude assignment is no longer needed for separation or when leaving Class B airspace. At that time, pilots must return to an altitude that conforms to VFR altitudes.

VFR aircraft operating in proximity to Class B airspace are cautioned against operating too closely to the boundaries, especially where the floor of the Class B airspace is 3,000 feet or less or where VFR cruise altitudes are at or near the floor of higher levels. Observance of this precaution will reduce the potential for encountering an aircraft operating at the altitudes of Class B floors. Additionally, VFR aircraft are encouraged to utilize the VFR Planning Chart as a tool for planning flight in proximity to Class B airspace. Charted VFR Flyway Planning Charts are published on the back of the existing VFR Terminal Area Charts.

CLASS C AIRSPACE

Generally, Class C airspace is that airspace from the surface to 4,000 feet above the airport elevation (charted in MSL) surrounding those airports that have an operational control tower, are serviced by a radar approach control, and that have a certain number of IFR operations or passenger enplanements per year. Although the configuration of each Class C airspace area is individually tailored, the airspace usually consists of a 5-NM radius core surface area that extends from the surface up to 4,000 feet above the airport elevation, and a 10-NM radius shelf area that extends from 1,200 feet to 4,000 feet above the airport elevation.

The outer area is normally a radius of 20NM, with some variations based on site-specific requirements. The outer area extends outward from the primary airport and extends from the lower limits of radar/radio coverage up to the ceiling of the approach control's delegated airspace, excluding the Class C airspace and other airspace as appropriate.

Class C airspace is charted on Sectional Charts, IFR Enroute Low Altitude, and Terminal Area Charts where appropriate. Operating rules within Class C airspace are a bit more relaxed than A and B airspace. There is no requirement for a level of pilot certification and required equipment includes only a two-way radio, and unless otherwise authorized by ATC, an operable radar beacon transponder with automatic altitude reporting equipment (Mode C).

Arrival

Arrival or flight through Class C airspace requires two-way radio communication which must be established with the ATC facility providing ATC services prior to entry and thereafter maintain those communications while in Class C airspace. Pilots of arriving aircraft should contact the Class C airspace ATC facility on the publicized frequency and give their position, altitude, radar beacon code, destination, and request Class C service. Radio contact should be initiated far enough from the Class C airspace boundary to preclude entering Class C airspace before two-way radio communications are established.

Note: If the controller responds to a radio call with, "(aircraft call sign) standby," radio communications have been established and the pilot can enter the class C airspace. If workload or traffic conditions prevent immediate provision of class C services, the controller will inform the pilot to remain outside the class C airspace until conditions permit the services to be provided.

It is important to understand that if the controller responds to the initial radio call without using the aircraft identification, radio communications have not been established and the pilot may not enter the class C airspace. Example:

[Aircraft call sign], "Remain outside the class Charlie airspace and standby," or "Aircraft calling Dulles approach control, standby." These calls from ATC indicate the pilot needs to remain outside Class C airspace and await class C services, which should come in a short time.

Departing

When departing from Class C airspace at a primary or satellite airport with an operating control tower, two-way radio communications must be established and maintained with the control tower, and thereafter as instructed by ATC while operating in the Class C airspace.

When departing a satellite airport without an operating control tower, the pilot needs to have two-way radio communications established with the ATC facility having jurisdiction over the Class C airspace as soon as practicable after departing from the satellite airport.

Unless otherwise authorized or required by ATC, no person may operate an aircraft at or below 2,500 feet above the surface within 4 nautical miles of the primary airport of a Class C airspace area at an indicated airspeed of more than 200 knots (230 mph). No problem for your basic Cessna 172. (FIG. 7-4.)

When two-way radio communications and radar contact are established, all participating VFR aircraft are sequenced to the primary airport and provided Class C services within the Class C airspace and the Outer Area. Aircraft are also provided basic radar services beyond the outer area on a workload-permitting basis. This can be terminated by the controller if workload dictates. I like it when I am departing a relatively quiet Class C airport and before I am 10 miles away I hear, "Radar service terminated, squawk VFR, frequency change approved." This is a sure sign the card game is about to commence.

Fig. 7-4. *A Cessna 340 cruises in Class C Airspace.*

Aircraft Separation

Aircraft separation is provided within the Class C airspace and the Outer Area after two-way radio communications and radar contact are established. VFR aircraft are separated from IFR aircraft within the Class C airspace by either visual separation or a 500 feet vertical separation except when operating beneath a heavy jet or target resolution on the scope.

Note: Separation and sequencing of VFR aircraft will be suspended in the event of a radar outage as this service is dependent on radar. The pilot will be advised that the service is not available and issued wind, runway information, and the time or place to contact the tower.

Pilot participation is voluntary within the outer area of Class C airspace and can be discontinued, within the outer area, at the pilot's request. Class C services will be provided in the outer area unless the pilot requests termination of the service.

Some facilities provide Class C service only during published hours. At other times, terminal IFR radar service will be provided. It is important to note that the communications and transponder requirements are dependant of the class of airspace established outside of the published hours.

In some locations Class C airspace may overlie the Class D surface area of a secondary airport. In order to allow that control tower to provide service to aircraft, portions of the overlapping Class C airspace may be procedurally excluded when the secondary airport tower is in operation. Aircraft operating in these procedurally excluded areas will only be provided airport traffic control services when in communication with the secondary airport tower.

Aircraft proceeding inbound to a satellite airport will be terminated at a sufficient distance to allow time to change to the appropriate tower or advisory frequency. Class C services to these aircraft will be discontinued when the aircraft is instructed to contact the tower or change to advisory frequency. Aircraft departing secondary controlled airports will not receive Class C services until they have been radar identified and two-way communications have been established with the Class C airspace facility.

These radar advisories should not be interpreted as relieving pilots of their responsibilities to see and avoid other traffic operating in basic VFR weather conditions, to adjust their operations and flight path as necessary to preclude serious wake encounters, to maintain appropriate terrain and obstruction clearance, or to remain in weather conditions equal to or better than the VFR minimums required by FAR Part 91.155. Approach control should be advised and a revised clearance or instruction obtained when compliance with an assigned route, heading and/or altitude is likely to compromise pilot responsibility with respect to terrain and obstruction clearance, vortex exposure, and weather minimums.

CLASS D AIRSPACE

Generally, Class D airspace is that airspace from the surface to 2,500 feet above the airport elevation (charted in MSL) surrounding those airports that have an operational control tower. The configuration of each Class D airspace area is individually tailored and when instrument procedures are published, the airspace will normally be designed to contain the procedures.

There is no specific rule regarding pilot certification in Class D airspace. In other words, student pilots are welcome, along with all the rest of us as long as they have two-way radio communication.

Arrival

In order to either arrive or fly through Class Delta airspace, two-way radio communication must be established with the ATC facility providing ATC services prior to entry and thereafter maintain those communications while in the Class D airspace. Pilots of arriving aircraft should contact the control tower on the publicized frequency and give their position, altitude, destination, and any request(s). Radio contact should be initiated far enough from the Class D airspace boundary to preclude entering the Class D airspace before two-way radio communications are established.

Sometimes I hear a pilot call the tower and report they are 5 miles out with the ATIS and want clearance into the Class D airspace. Whoops. By the time the tower answers, they are already in the Class D airspace. God forbid, if the tower were busy, that pilot might just fly on through the Class D airspace before they receive clearance. Plan ahead!

If the controller responds to a radio call with, "aircraft call sign, standby," radio communication has been established and the pilot can enter the Class D airspace. If workload or traffic conditions prevent immediate entry into Class D airspace the controller will inform the pilot to remain outside the Class D airspace until such time as the controller has time to work them in.

If the controller responds without using the aircraft call sign, two-way communication has not been established, and the aircraft may not enter the Class D airspace. Example: "Aircraft calling the tower, stand by."

Departure

When departing from the primary or satellite airport with an operating control tower, two-way radio communications must be established and maintained with the control tower, and thereafter as instructed by ATC while operating in the Class D airspace. Operations from a satellite airport without an operating control tower also requires two-way radio communications be established as soon as practicable after departing with the ATC facility having jurisdiction over the Class D airspace.

Unless otherwise authorized or required by ATC, no person may operate an aircraft at or below 2,500 feet above the surface within 4 nautical miles of the primary airport of a Class D airspace area at an indicated airspeed of more than 200 knots, which is 230 mph (see FIGS. 7-5 and 7-6).

Class D airspace areas are depicted on Sectional and Terminal charts with blue segmented lines, and on IFR EnRoute Low Altitude Charts with a boxed [D]. Arrival extensions for instrument approach procedures may be Class D or Class E airspace. As a general rule, if all extensions are 2 miles or less, they remain part of the Class D surface area. However, if any one extension is greater than 2 miles, then all extensions become Class E.

Fig. 7-5. *A Cessna 185 climbs through Class D Airspace*

Fig. 7-6. *This Cessna Skymaster is among a diverse group of aircraft often found in Class D Airspace.*

It is important to remember there is no radar at Class D airports, therefore no separation is provided for VFR Aircraft. It is up to the pilot-in-command to see and be seen.

At the Class D airports, there are no specific requirements concerning equipment, pilot certification, or for Mode C operations capability. Even two-way radio communication may be by-passed with a call to the tower about an hour prior to arrival.

CLASS E AIRSPACE

Class E airspace is all other controlled airspace not referred to as either A, B, C, or D airspace. There is no certification nor equipment requirements for operations in Class E airspace. Class E airspace below 14,500 feet MSL is charted on Sectional, Terminal, World, and IFR EnRoute Low Altitude charts. Except for reaching 18,000 feet MSL, the floor of Class A airspace, Class E airspace has no defined vertical limit but rather it extends upward from either the surface or a designated altitude to the overlying or adjacent controlled airspace.

Class E airspace is designed in several different ways to augment the purpose at a given site. When Class E airspace is designated as a surface area for an airport, the airspace will be configured to contain all pertinent instrument procedures. When designated as an extension to a surface area, there are Class E airspace areas that serve as extensions to Class B, Class C, and Class D surface areas designated for an airport. Such airspace provides controlled airspace to contain standard instrument approach procedures without imposing a communications requirement on pilots operating under VFR.

There are Class E airspace areas beginning at either 700 or 1,200 feet AGL used to transition to/from the terminal or enroute environments around smaller airports. These transition areas are designated on Sectional charts by a magenta colored ring encircling the airport with a diameter of 10 miles.

EnRoute Domestic Areas are Class E airspace areas that extend upward from a specified altitude and provide controlled airspace in those areas where there is a requirement to provide IFR enroute ATC services but the Federal airway system is inadequate.

The Federal Airways, also known as "Victor" airways are Class E airspace areas and, unless otherwise specified, extend upward from 1,200 feet to, but not including, 18,000 feet MSL. The colored airways are Green, Red, Amber, and Blue. The VOR airways are classified as Domestic, Alaskan, and Hawaiian.

Offshore Airspace Areas are Class E airspace areas that extend upward from a specified altitude to, but not including, 18,000 feet MSL and are designated as offshore airspace areas. These areas provide controlled airspace beyond 12 miles from the coast of the United States in those areas where there is a requirement to provide IFR enroute ATC services and within which the United States is applying domestic procedures.

CLASS G AIRSPACE

Much of the concern about our airspace system stems from a widespread misunderstanding about what causes controlled airspace to be controlled. Airspace is controlled by only two things: weather minimums that force you to meet VFR requirements or an area that needs, due to traffic loads, the special guidance of Air Traffic Control (ATC).

Our uncontrolled airspace (Class G) is that portion of our air that has not been designated as either class A, B, C, D, or E airspace. The vast majority of this Class G airspace lies below transition areas and has a maximum altitude of 1,200 feet AGL. This uncontrolled airspace, for the most part, lies below the 1,200-foot AGL ceiling between airports that have no FAA controlling facilities. In those uncontrolled areas, you may fly VFR during daylight hours, to anywhere you desire, as long as you have at least 1-mile visibility and remain clear of the clouds. At night, your visibility requirement rises to three miles visibility along with standard cloud clearances of 500 feet below, 1,000 feet above, and 2,000 feet horizontally.

If you find Class G airspace in an area that is above the 1,200-feet AGL level but below 10,000 feet MSL, you then have to have at least 1 mile of visibility and fly no closer than 500 feet below, 1,000 feet above, and at least 2,000 feet horizontally from any clouds. At night, the cloud clearances remain the same, but the visibility requirement increases to three miles.

In Class G airspace that is both more than 1,200 feet above the surface and at or above 10,000 feet MSL, you must have at least five miles visibility and remain 1,000 feet above and below the clouds as well as one mile horizontally.

8
Emergency Procedures

I WAS ON FINAL APPROACH WHEN IT HAPPENED. WITH NO WARNING, the only engine I had quit cold. The familiar hum of the engine ceased, the rpm wound down and the propeller stopped. As I looked through the windscreen at the motionless propeller, time also seemed to freeze. I was only 20 to 25 seconds from touchdown, yet I had inadvertently placed my life in jeopardy. I had a very serious problem. I was going down and I could instantly see I would not make the runway.

The day started as a combination of work and play. I was to test fly a new airplane, an experimental airplane, and I was high with anticipation. The airplane was beautiful, and the rain, which had been relentless for two days, had finally let up. I was ready to fly.

The day unfolded uneventfully, although fun. The brisk, 20 to 30 knot northwesterly winds were a nuisance, but the airplane flew flawlessly as I put it through its paces for most of the morning. After lunch, I was running a series of speed and fuel consumption tests and had to land frequently to assess items and meter fuel.

The runway was lined up north/south, clear on the north, but bounded on the south by a very high tree line that had their roots in the Ohio River. Man, were they tall. As the winds were primarily from the northwest, I had been taking off and landing to the north which necessitated that my final approach to the crosswind landing be over the tall tree line.

Chapter Eight

On this particular approach, I was on final carrying a bit of power, slow-flying the airplane down an invisible line over the trees in order to land as short as possible. This was necessary since the 100-foot tall trees shortened the effective runway length appreciably.

I was at about 400 feet above the ground, crabbed into the strong northwesterly crosswind, and fat, dumb, and happy as they say, when it happened. In less than a heartbeat, the engine stopped running. Instantly, I glanced around and knew I was in deep trouble. The powered approaches I had been making over the trees had been as safe as safe could be. As long as the engine kept running. Now I had some decisions to make.

My first thought as the airplane began to sink beneath the tops of the trees was to dive, pick up speed, zoom over the trees, point the nose down, and glide in to land. Not a great idea.

In less time than it took to think it, I discarded this idea as placing all my eggs in one basket. This was a true all or nothing idea. Either it would be a roaring success, I would make it, and we would all laugh about it over a beer, or I would surely die. Not good enough odds.

As I checked the fuel selector and primer to see all was where it should be, I glanced to my left and saw trees as far as I could see. Which wasn't very far as I was now down to about 200 feet and sinking like the proverbial rock. Well, I thought, if I turn northwest, at least I can slow my groundspeed down to the point that when I hit the trees, I might survive it. Better, but not good enough.

To my right lay a newly sprouted wheat field, long, light green, and downwind. Downwind? I remember thinking that even with the 25-knot tailwind, the field was long enough that I could get it down, but could I get it stopped? It was about this time, as I turned toward the wheat field, it dawned on me. I wouldn't have any problem stopping. The mud in the newly planted wheat field would have to be about 6 inches deep. Oh, boy!

As I neared the wheat field, I cursed the rain, the mud, and my rotten luck. But, I was out of choices and I still had to go through with this. I shut off the mags and the fuel, and switched off the master switch. At least this should prevent the airplane from burning.

I turned my full attention to the landing, which seemed to take forever in my downwind state. I lowered the manual flaps to the full down position and waited. As I neared the muddy field, I pulled my seat and shoulder straps tighter and prepared for the inevitable sight I knew had to be coming, the noseover in the deep mud and my inverted ending.

The airplane touched down ever so lightly in a full stall and I held the stick back as far as I could hold it. For a few fleeting seconds, I thought I just might get it stopped as the airplane slowed rapidly in the deep mud. The mud was slamming against the aircraft with such fury that it sounded like it would come through the skin.

Then, I saw my Waterloo coming in the form of a small ditch running perpendicular across in front of me. I held the stick full back and waited. When the mains hit the ditch, the airplane nosed over very quickly, slammed upside down in the mud, and everything went dark.

I quickly assessed my situation. The first thing I noticed was my head was wet and that it was absolutely pitch dark where I was. I figured that as the airplane inverted, the bubble canopy had shattered and the airplane's fuselage was imbedded firmly, though upside down, in the mud. Okay, dummy, of course your head is wet, it's in the mud. And this would also explain the absence of any light.

That's okay, I thought, I'm awake and I don't hurt, but how can I get out? I put my hands up (down) and felt all around only to feel mud, glass, and water. And I smelled fuel. Fuel? Oh, boy! Here I am upside down with the canopy smashed to pieces, the cockpit shoved into the mud, I can't get out, and I smell fuel. Swell.

As I had mumbled a fervent prayer to my Lord to please help me out, I heard a sound that almost stopped my heart. Tick, tick, tick, tick. It was the fuel pump! This meant the master switch was on and there was power emanating from the electrical system which was producing enough power to run the fuel pump. This, coupled with fuel running everywhere, meant that I had to eliminate the source of the spark or risk burning to death.

In the total darkness, I felt down (up) to approximately where the master switch was located, fortunately found it in a second, shut it off, and relished in the absolute silence. (I later discovered a cut on my knee made when the airplane went inverted that was probably caused by hitting the master switch and turning it back on.)

Thank God, help arrived quickly, the airplane was lifted enough so I could crawl out from under it, and I was free. Boy, the light and fresh air looked and smelled good!

I came through the incident with but a few aches and a couple of small cuts. The airplane wasn't so lucky. It took quite a few more hours to fix the plane than it did to fix the pilot. But when you're the pilot, that's good news.

So, what did I learn from replaying this 30 second slice of my life about 15,000 times? I learned once again that I am mortal and that life can end in a moment. I also learned that a new airplane does not necessarily come without flaws. But mostly I am contented that I did about all I could have done, given the circumstances, and am happy I performed in the appropriate manner. As you have read, I had several opportunities to err and but one possible choice to survive, God willing. The ability to sort through three or four choices and pick the most survivable is what I have trained for all my life. I'm glad to say it worked.

Nothing actually prepares a person for the onset of a full-blown in-flight emergency, but proper training and practice usually will win out provided one does not allow panic to overcome common sense.

For some rather strange reason, immediately after this incident, I found the whole thing rather funny. Someone had called the police and fire departments and about six cars and ambulances and fire trucks pulled in at the crash sight within about 15 minutes. I don't know if I was embarrassed or happy to be alive or what, but when a rather good-looking female EMT came up to me and without saying anything, began running her hands all over my body in search of some unseen and unfelt wounds, I asked her if she was with the fire department or just taking advantage of an opportunity?

ENGINE FAILURE

Emergencies come in all shapes and sizes. They can range from fire to complete engine failure, loss of an aircraft component, radio failure, letting your maps blow out the window, or almost anything else you can dream up. It would be impossible to try to cover all the different possibilities here in this chapter. However, they all have one thing in common. They all need action. The right action. In most emergency situations, the pilot must be conditioned to respond quickly and correctly.

Some emergencies are much more serious than others. And what seems to be an emergency to one person might be only an annoyance to someone else. The defining element of each emergency is whether it seems to be an emergency to the pilot. So if you believe you have an emergency, then by golly it is! At least to you.

Sometimes, as in an engine failure at high altitude, you have plenty of time to get organized and proceed accordingly. Other times you must make split-second decisions. Let's look at the problem that usually comes to mind first for most pilots—engine failure. This particular emergency can be very dangerous or no big deal, depending on where it happens and the pilot's readiness, training, and reaction to the engine failure. Whether the failure is total or only a partial power loss also makes quite a difference.

Generally speaking, there are three distinct actions you should complete, in order, if you suffer power loss:

- Set up a glide.
- Make a thorough cockpit check.
- If you cannot get a restart, turn into the wind (or crosswind) and land on the best available surface. Never land downwind if you can help it.

Glide

Let's go through the reasoning behind these three actions and determine why they should come one at a time, in this exact order. If you are anywhere other than on the ground (which is the best place to experience a power loss) what is the most important thing you have going for you? *Altitude.* Altitude gives you time. It gives you time to think, act, and call to ask for help or just to let someone know you are having problems. It gives you time to ready yourself and your aircraft for a possible emergency landing and time to pick the best possible spot to put down. There is even time to attempt a restart. Altitude buys you time, so in order to obtain the most time, instead of letting the aircraft descend rapidly in a cruise descent, which eats up huge chunks of altitude, set up a glide using the manufacturer's recommended best glide speed, save that altitude, and use it to your advantage. That's number one.

Cockpit Check

Number two is a thorough cockpit check. Having set up a glide, and assuming you are at a reasonably high altitude, you should have time to investigate and maybe find and correct the cause of the power loss. Maybe you need carb heat to get rid of a small amount of carburetor ice, or maybe you need to switch to another fuel tank (one with some fuel in it).

You would be surprised at the number of very major accidents that have been caused by some very minor problems. Many of these problems could have been overcome with a proper cockpit check.

When you perform your cockpit check, you should go to the most likely causes of power loss such as carb heat, fuel valve, mixture, mags, and the primer. However, don't go for them in a random pattern that might cause you to overlook something very important. Do it in a systematic fashion. If time permits, use your emergency checklist. Chances are that your mind will be in some degree of shock from the sudden loss of power but with a written checklist you are less apt to forget something.

I teach my students a method to use in an emergency if there isn't time or they can't locate the checklist. I believe it gets the job done quickly and thoroughly.

As you are setting up a glide in a Cessna 152 (FIG. 8-1), you are cranking in some trim to stabilize the airspeed at the best glide speed. Since your hand is already on the trim tab, it is a very easy and short motion to drop your hand almost straight down from the trim tab to check the fuel selector. Then, bring your hand up and slightly to the right and check the mixture for full rich. The rest of the motion is directly to your left as you progress across the lower panel and check the carb heat (which I would have pulled on the very first thing), the mags (which should be checked one at a time), and the primer (which should be in and locked).

Fig. 8-1. *Cessna 152 panel.*

This method covers the main points and allows you to save the time you would otherwise lose looking for the checklist. It also can be completed very quickly. It is especially useful in a low-altitude emergency situation where you just don't have time to fumble for a written checklist (See FIG. 8-2). At the risk of repeating myself, I want to make one point very clear: *If you have the time and altitude, use the written checklist.*

Turn into the Wind

Number three is to turn into the wind, or crosswind, and find the best place you can to put it down. This step should be taken if all else fails and you cannot get a restart. In this event, land as slowly as possible and with as much control as possible. The cardinal rule of aviation emergencies is, "If you know you are going to have to land, or crash, go in with as much control as possible." Then, you will have some control over your destiny.

EMERGENCY LANDINGS

If you have the unpleasant experience of having to make an unscheduled landing in an emergency, how do you look for the best place to land? I have a set of rules I teach my students for finding the best place to make an emergency landing.

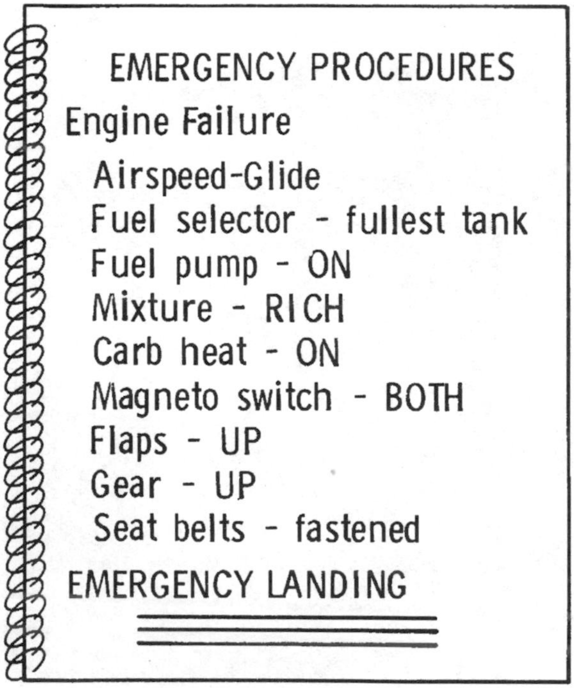

```
EMERGENCY PROCEDURES
Engine Failure
    Airspeed-Glide
    Fuel selector - fullest tank
    Fuel pump - ON
    Mixture - RICH
    Carb heat - ON
    Magneto switch - BOTH
    Flaps - UP
    Gear - UP
    Seat belts - fastened
EMERGENCY LANDING
```

Fig. 8-2. *Emergency checklist. Don't leave home without it!*

The first choice and the most obvious is an airport. Why put it in a field right beside a runway? (Don't laugh; it's been done.)

The second choice is a road, vacant of traffic and void of power lines. It is your responsibility to take every option open to you to prevent any injury to innocent people on the ground.

My third choice is a pasture, either hay or wheat, freshly cut or early in the season so it will be short, reducing the chances of a noseover after touchdown. All the rest follow in a rather random order as none of them are very conducive to a successful emergency landing.

If you ever know that you are going to have to land in a field of any kind, you will be far better off to land *with* the rows rather than *across* them. This rule also applies to plowed ground. Anything that decreases your chances of remaining upright increases your chances of bodily injury. Other than innocent people on the ground, the most important thing for you to save is yourself.

Here is another example from NTSB file # 3-3685. Near Cherryfield, Maine, an ATP-rated pilot flying a Cessna 150 suffered a total powerplant failure and had to make an emergency landing. In the course of the landing, the gear collapsed, causing substantial damage to the aircraft. The injuries are listed as either minor or none. The probable causes and factors list contains powerplant failure for undetermined reasons, terrain-rough/uneven, overload failure, and gear collapse.

This pilot evidently did a good job of handling this emergency because there were only minor injuries. The point is that aircraft are made to land on a hard, smooth surface. For the most part, aircraft landing gear cannot withstand landing on rough, uneven surfaces (FIG. 8-3).

POWER FAILURE

If you suffer a power failure during the takeoff roll, abort the takeoff. Reduce power (if you have any) to idle and stop. Don't try to struggle into the sky with an aircraft that is not developing full power. It's suicide.

If you suffer a power failure after liftoff, then there are several significant variables that must be taken into account that will be the deciding factors on what moves you need to pursue. These items include the amount of runway remaining, altitude at the time of failure, and possible obstructions in the flight path. A general rule recommended by the FAA and taught by most flight instructors is that if you are not at least 500 feet above the surface, continue on straight, take whatever comes, and land as slowly as possible. Most pilots cannot safely perform the 180-degree plus turn and return to land safely below 500 feet. Many have tried; few have made it. Most stall during the turn or run out of airspeed, altitude, and ideas, all at the same time. The result? The probability is a much more severe crash than likely would have occurred had they continued straight ahead and landed the aircraft as slowly and with as much control as possible.

ZONE OF DECISION

If you are between 400 and 1,000 feet above the ground, the area I call the *zone of decision* you will still have to take into consideration the factors of altitude, runway avail-

Fig. 8-3. *The field may look good from the air, but sometimes things don't always go as planned.*

ability, obstructions, and safe landing areas. It is my personal belief that this decision area is the most dangerous place to suffer a power loss. There are so many things you could do that making the correct decision is sometimes very difficult. Your first question is, where can you land with the best chance of doing so safely? And that choice is probably where you should go. Forget the what ifs and the maybes. Go to the spot you know offers the best chance for a safe landing.

In this zone of decision, you should have time to set up your glide, make a cockpit check, and providing you cannot obtain a restart, pick the most suitable place to put down. You have to practice your emergency techniques until they are automatic reflex actions. Emergency techniques should be almost mechanical. They have to be. Sometimes you just don't have time to get out the book.

I have had the occasion to make several emergency landings during my time in aviation; fortunately, most were of the precautionary type. Two were not. And one, which I will remember until the end of time was a combination takeoff, emergency, and landing.

On a hot summer day several years ago, two other instructors and I were going someplace in a Piper Apache. I don't remember where we were going because we never got there. In fact, we didn't even make it to the departure end of the runway.

An Apache is a four-place, twin-engine, and relatively low-powered aircraft. It was one of the first production twins affordable to pilots of a near-average financial status.

The aircraft was a pretty good plane as long as both engines were turning. It was not known for its single-engine performance.

I was elected to fly that day. With everything checked and ready, we started our take-off roll on runway 27. The aircraft accelerated a tad slowly, mostly because we were at near gross weight and it was a very hot day. Our runway was 5,200 feet long and I suppose we had used about 1,200 feet before we reached our rotation speed, safely above V_{mc} (velocity minimum control), rotated, and were off.

At a height of about 30 feet above the runway, the right engine quit as though it had fallen off of the aircraft. It didn't sputter, cough, or anything else. It simply came to an abrupt halt. Before I could react, the nose had swung about 20 degrees to the right of our runway centerline and we were headed for the boonies, and not climbing. I immediately chopped the power to both engines, swung the Apache back to the runway, hit full flaps, landed, put the flaps back up for better braking, and stood on the binders. We stopped with little runway to spare.

Had I been on a shorter runway, we would have been committed to fly on and try to nurse the Apache to enough altitude to allow us to come back and land. But we had a full mile of hard-surfaced runway to work with, so I had to make a decision. I opted to chop the power and land because I thought it would be the safest thing to do. And I didn't have all day to make that decision, nor did I bother to ask anyone else what they thought we should do. It was reflex. Reflex that came from some very good dual instruction in multiengine emergency techniques and many hours of practice and thought. The result was no damage to the aircraft or its occupants, although there was some good-natured talk of cleaning the seats after it was all over and done.

It seems our problem was a faulty fuel selector valve. Although it appeared to be in the right main position, it was indeed off. It just goes to show that you have to be ready at all times.

This example brings me to a point I teach to my students and try to remember myself I call it the "What would I do if?" problem. When you are flying alone, taking off, landing, or whatever, look around and ask yourself, "What would I do if?" Try it. It's good fun and will make you aware of things you have never dreamed. Sometimes I find myself at home, sitting in my chair, thinking of a new situation, and asking myself, "What would I do if?" It's almost as good as getting some instruction.

If you experience a power failure at an altitude above 1,000 feet, follow the three rules talked about previously. You should have time to utilize your emergency checklist and be certain you have checked every possible emergency procedure before making the decision to land. It is especially true if you happen to have the misfortune of being over some rather hostile terrain (Fig. 8-4). Above all, take your time and try not to panic. Panic has probably caused some routine emergencies to terminate with unnecessarily severe consequences.

Flight instructors know that a student who reacts in a poor or erratic manner when practicing simulated emergency landings will most likely have problems if the emergency ever becomes a reality. All of us have our heart rate quicken when confronted by a threat. Some people handle this stress much better than others. Actually, this quickening

Fig. 8-4. *This Cessna 185 would have a very hard time finding a suitable landing area in this rugged terrain.*

of the heart rate and the increased flow of adrenaline is our body's way of getting us geared up to meet the challenge. Some pilots react in a calm, efficient manner while others seem to use the extra blood and adrenaline in order to do *something*, whether right or wrong. They panic. All their emotions are so caught up in the instinct for survival that they forget their training, lose their thought process, and usually wind up acting incorrectly. The results are often fatal.

A pilot should be able to react within the scope of his training and experience in a calm, efficient manner when faced with an emergency. This action is the mark of a professional. He doesn't let the emergency handle him. He handles the emergency. Once you give up being the pilot-in-command and become what amounts to a passenger, you are most certainly in deep, deep trouble.

There are many other types of emergencies you might encounter during your aviation career. Most of you, however, will probably never come into contact with an emergency of any sort because flying is safe. And it's becoming safer every day. Pilots are being better trained and aircraft and their powerplants are becoming more reliable. Although there are more people flying more aircraft now than ever before, the percentage of accidents versus hours flown is declining.

ACCURACY LANDINGS

Accuracy landings are included with the emergency procedures because the main purpose for accuracy landings is to land your aircraft on a given spot if the need should arise.

They are a very important part of every pilot's training. Knowing you have the ability to land an aircraft on a desired spot, with or without power, is a very comforting feeling. Although accuracy landings are fun to practice and should be on the private pilot checkride, they have a much more serious purpose. They are invaluable in an actual emergency situation.

The most common procedure for practicing accuracy landings is to pick a spot on the runway and then attempt to land on or within 200 feet beyond the spot. Any landing made short of this spot must be considered a failure. In an actual emergency, if you fail to make the field by even a few feet, the chances of your striking a fence or some other object are greatly increased. For this reason, I never have my students use the end of the runway because it leaves no margin for error. I usually have them use a centerline stripe that is at least a couple of hundred yards down the runway. If you practice on a sod field, you can use anything you can find that shows itself well from the air. It could be a spot that has a different shade to it or perhaps a slight rise or indentation on the runway.

To begin practice on accuracy landings, enter downwind at the normal pattern altitude and pick a spot on the runway that is fairly prominent. When your aircraft reaches a point adjacent to the spot, reduce power to idle. The pattern should remain rectangular and as normal as possible. The use of flaps, slips, or slight S-turning is encouraged as long as you don't get too carried away and start using dangerous maneuvers to arrive at the spot.

It's best to try to turn onto final approach a little high, if possible, because you can always lose altitude by the methods just mentioned. You should be careful that you don't stay so high that you might have a problem with an overshoot. Remember, this is an emergency procedure. In real life you will probably get only one chance. On the other hand, if you are too low, there is no way on earth to get your aircraft onto the runway without power.

Always remember to aim a little short of the spot to allow for the float during the flare. The method I teach is the old *moving spot technique*. When you turn onto final, watch the spot of intended landing. If the spot appears to move toward you, you are too high. If it appears to move away from you, you are too low. Of course, if you are way too low or high, you might not make the runway at all, or overshoot in the case of the latter. It has to be within reason and it also depends on how far out you are from touchdown, airspeed, flap setting, etc. This method is as foolproof as any I have ever seen, and with a little practice you can land on a given spot with little difficulty.

Accuracy landings should be practiced with and without power, in full flap and no flap configurations. They should be practiced until you are confident that you can land on a given spot from any altitude and power setting. Then, continue to practice them often so you don't lose your touch. The knowledge that you can put your aircraft down on a given spot from any altitude at any time is one of great comfort and security.

SPIRAL

Now that we have discussed emergency procedures and accuracy landings, I think it would be beneficial for you to learn one more time-proven method that you can utilize to get down from a relatively high altitude while remaining over your point of intended landing. This method is called a *spiral.*

Spirals are useful in your training to help you learn to maintain orientation during prolonged descending turns. They help increase your ability to get your head out of the cockpit and still control the airspeed and bank. In the event of an actual emergency, the spiral is probably the best way to lose altitude and remain close to the point of intended landing. A spiral is also useful when coming down through a hole in the clouds. It prevents the possibility of illegal IFR flight. I know many pilots who have used this method to come down after getting so engrossed in practice that they failed to notice that the broken layer of clouds they had at the start of the flight had become nearly overcast.

Spirals are essentially a high-altitude emergency technique (FIG. 8-5). Whether spiraling about a point to an emergency landing or coming down through a broken cloud layer, the spiral usually indicates the need for some type of prompt action.

As far as I am concerned, spirals are divided into two categories. The first is used in the instance I just mentioned. It is useful when coming down through a hole in the clouds

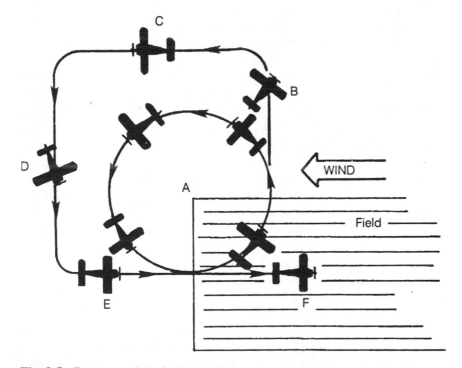

Fig. 8-5. *Emergency Spiral; A) Spiral about the downwind corner of the downwind side of the field to stay close to your point of intended landing; B) Break out of the spiral several hundred feet above a normal pattern altitude; C) Turn downwind slightly higher than normal; D) Turn base slightly higher than normal; E) Turn to final slightly higher than normal. Use slips or flaps, as necessary, to land on intended spot; and F) Land short using correct technique (soft-field, rough field, etc.).*

or whenever a rapid descent is called for. Since this spiral does not include ground tracking, your airspeed and bank control are your main concerns.

To begin your practice for this particular spiral, attain a fairly high altitude, close the throttle, set up a normal glide, and perform a cockpit check. Although this might not be an emergency situation, it is very important to maintain these habits. After clearing the areas below, begin the spiral using about 50 degrees of bank. In this particular spiral, since ground track is not the major purpose, closely monitor the airspeed and bank control through the desired number of turns. As a good safety practice, you should not spiral down lower than 1,500 feet above the ground. Continue your practice until you can perform at least three turns and keep your airspeed and bank constant.

The second type of spiral is more in the category of ground tracking. Incorporate the methods utilized in the previous spiral, but use a point on the ground about which you will spiral. It then becomes only a matter of doing a turn about a point in a gliding turn. Vary your bank as necessary to maintain a constant distance from the pivotal point. Upon reaching pattern altitude or slightly above, leave the spiral and enter a normal traffic pattern to the place of intended landing.

The use of the traffic pattern is wise because the more familiar the situation seems, the easier it will be for you to land on the intended spot. Most people have more problems landing out of a straight-in approach than from a normal traffic pattern. If you add a margin for human error due to the fact that an emergency situation might cause less than superior performance, I think you will see the need for establishing a normal traffic pattern.

I teach my students to spiral about the downwind corner on the downwind side of the landing area they are trying to get into. It places them in the best possible position to execute a near normal traffic pattern and provides a constant to strive for which is desirous in any emergency situation.

9
Multiengine Flight

TRANSITIONING TO COMPLEX AIRPLANES

Multiengine aircraft can be deceiving. To the uninitiated, the multiengine aircraft appears twice as safe as its single-engine counterparts. I have often heard it said that with two engines a pilot has twice the insurance, twice the speed, and twice the redundancy that a single-engine pilot has. They also have twice the chance of having an engine fail.

And herein lies the tale. The multiengine aircraft is a thing of wonder: an aircraft that usually carries more weight, flies faster and farther, and offers the redundant safety of a spare engine. But all of these good points go out the window very quickly when an engine is lost on a multiengine aircraft. A multiengine aircraft operating on one engine can turn into a beast of a different color. And it can be a killer. In the hands of the unskilled or the careless, the multiengine aircraft operating on one engine is tantamount to playing Russian roulette with about 5 bullets in the chamber.

It's not that everyone who flies a multiengine aircraft is playing with fire. Quite the contrary. Treated with respect and not pushed beyond the laws of physics, a multiengine aircraft can be a very safe mode of transportation.

This chapter is devoted to the factors associated with, and the basic operating practices applicable to, transitioning into multiengine airplanes. As I mentioned, these airplanes have significantly different flight characteristics, performance capabilities, and operating procedures from those airplanes which the single-engine pilot has previously flown. However, accident records indicate that some pilots take unnecessary risks when they attempt to fly a different type of airplane without familiarizing themselves with its peculiarities, limitations, and systems. A knowledge and observance of the basic practices discussed in this chapter will hopefully prevent unnecessary accidents.

The increasing complexity of multiengine airplanes dictates the importance of a thorough checkout for pilots who change from one make or model airplane to another with which they are not familiar. The similarity of the operating controls in most airplanes leads many persons to believe that full pilot competency can be carried from one type of airplane to another, regardless of its weight, speed, performance characteristics, and limitations, or how many engines they possess (see FIG. 9-1).

The importance of acquiring a thorough knowledge of an unfamiliar airplane and the inefficiency of trial-and-error methods of learning to fly that airplane have been well established. So the pilot desiring to add a multiengine rating to their certificate will need some good dual instruction prior to applying for the multiengine checkride. In order to do this efficiently they must engage the services of a flight instructor.

Naturally, the pilot should obtain the services of a flight instructor who is fully qualified in the airplane concerned. The flight instructor should not only be well qualified in the airplane to be used, but also should be capable of communicating effectively to the pilot the techniques essential for the safe operation of the airplane. In other words, your brother-in-law may own a nice twin and be a great guy, but he probably isn't a flight instructor.

The pilot transitioning to multiengine should study and understand the airplane's flight and operations manual. A thorough understanding of the fuel system, electrical

Fig. 9-1. *Moving up to a twin-engine aircraft like this Cessna Corsair only requires motivation, time, and money.*

and/or hydraulic system, empty and maximum allowable weights, loading schedule, normal and emergency landing gear and flap operations, and preflight inspection procedures, is essential.

It is also very important that the transitioning pilot study the engine and flight controls, engine and flight instruments, fuel selector controls, wing flaps and landing gear controls and indicators, and radio equipment until proficient enough to pass a blindfold cockpit check in the airplane in which qualification is sought.

In order to really learn the operating characteristics of the new aircraft, do not limit familiarization flights to the mere practice of normal takeoffs and landings. It is extremely important to learn the "V-speeds," and become thoroughly familiar with the stall performance, minimum controllability characteristics, maximum performance techniques, and all pertinent emergency procedures, as well as all normal operating procedures.

The transition from training type single-engine airplanes to larger and faster multi-engine airplanes may be the pilot's first experience in airplanes equipped with a constant-speed propeller, a retractable landing gear, and wing flaps. And all airplanes having a constant-speed propeller require that the pilot have a thorough understanding of the need for proper combinations of manifold pressure (MP) and propeller revolutions per minute (RPM) which are prescribed in the airplane manufacturer's manuals.

When transitioning to high-performance or complex airplanes, the pilot must he cautioned not to exceed the specified combinations of power settings since the engine can be damaged by using excessively high MP with a low engine RPM. If this situation were to occur, the BMEP (Brake Mean Effective Pressure) might be exceeded. BMEP refers to the average internal pressure exerted upon the cylinder walls and pistons in the combustion chamber during the power stroke. To preclude excessive stress on the engine when increasing power, the pilot should first move the propeller levers forward, increasing the engine RPM, and then advance the throttles. When reducing power, the throttles should be retarded first, followed by the propeller controls.

Before applying full power during takeoff, the propeller controls should be placed full forward (high RPM, low pitch) to protect the engine from excessive internal pressures. After takeoff, the MP should be reduced first to normal climb settings, followed by the propeller controls. This procedure should never be performed in the reverse order.

On the approach to a landing, when the airplane is committed to a landing, the propeller controls should be placed to a high RPM position so that should it be necessary to advance the power for a go-around, the propeller will have the correct pitch for the maximum thrust.

The instructor should include in the checkout of any aircraft, at least a demonstration of takeoffs and landings and in-flight maneuvers with the airplane fully loaded. Most four-place and larger airplanes handle quite differently when loaded to near-maximum gross weight, as compared with operations when lightly loaded. Weight and balance should also be figured for various loading conditions.

It is very important that the transitioning pilot readily accept the flight instructor's evaluation of performance during the checkout process. It is inadvisable to consider oneself qualified to accept responsibility for the airplane before the checkout is completed; half a checkout may prove more dangerous than none at all.

NORMAL MULTIENGINE PROCEDURES

Today's technology has produced safe, efficient, and modern multiengine airplanes. Their utility and acceptance have more than fulfilled expectations of their builders. As a result of this rapid development and increasing use, many pilots have found it desirable to make the transition from single-engine airplanes to those with two or more engines and complex equipment. Good basic flying habits formed during earlier training, and carried forward to these new sophisticated airplanes, will make this transition relatively easy, but only if the transition is properly directed.

The increased complexity of multiengine airplanes requires a very systematic approach to preflight inspection prior to entering the cockpit, and a dedicated use of checklists for all operations.

Preflight visual inspections of the exterior of the airplane should be conducted in accordance with the manufacturer's operating manual. The procedures set up in the flight manuals usually provide for a comprehensive inspection, item by item, as with all aircraft, irrespective of their complexity. The transitioning pilot should have a thorough briefing and understanding of each item in the preflight inspection as it may be quite different from the pilot's former aircraft.

Checklists

As with any aircraft, the multiengine aircraft comes replete with checklists. And they are usually more complex and lengthy than the one's the pilot has been used to. The multiengine checklist is divided into common areas of operation, as are all airplanes, but will have a few areas single-engine airplanes do not have.

The multiengine aircraft also characteristically have more controls, switches, dials and instruments than its single-engine brethren. Some of these items carry a more serious penalty if not properly positioned before flight. This teaches the would-be multiengine pilot to carefully consider the results of their actions, or inactions. For instance, the multiengine pilot has two of everything, including fuel selector knobs, mixture controls, etc., which give the pilot twice the opportunity to miss an important item from the checklist.

If you have the luxury of having a front seat passenger who can read without missing a line, you might consider having them read you the checklist. But then you are placing the direct responsibility for the safe conduct of your flight in the hands of a layman. Although it might be them who misses an important item, everyone on board might pay the penalty.

The pilot of the multiengine aircraft should develop the habit of always actually touching the control or device and repeating the instrument reading or proscribed control position in question, under the careful observation of the pilot calling out the items on the checklist. Even when no copilot is present, the pilot should form the habit of touching, pointing to, or operating each item as it is read from the checklist.

In the event of an in-flight emergency, the pilot should be sufficiently familiar with emergency procedures to take immediate action instinctively to prevent more serious situations. However, as soon as circumstances permit, the emergency checklist should be reviewed to ensure that all required items have been checked.

TAXI PROCEDURES

The basic principles of taxiing which apply to single-engine airplanes are generally applicable to multiengine airplanes. Although ground operation of multiengine airplanes may differ in some respects from the operation of single-engine airplanes, the taxiing procedures also vary somewhat between those airplanes with a nosewheel and those with a tailwheel-type landing gear. With either of these landing gear arrangements, the difference in taxiing multiengine airplanes that is most obvious to a transitioning pilot is the capability of using asymmetrical power between individual engines to assist in directional control (see FIG. 9-2).

Tailwheel-type multiengine airplanes are often equipped with tailwheel locks which can be used to advantage for taxiing in a straight line, especially in a crosswind. The tendency to weathervane can also be neutralized to a great extent in these airplanes by using more power on the upwind engine, with the tailwheel lock engaged and the brakes used as necessary.

Braking

On nosewheel-type multiengine airplanes, the brakes and throttles are used mainly to control the momentum, and steering is done principally with the steerable nosewheel.

Fig. 9-2. *Asymmetrical, or unequal power is often used on multiengine aircraft to aid in taxi procedures.*

The steerable nosewheel is usually actuated by the rudder pedals, or in some larger airplanes by a separate hand-operated steering mechanism.

Brakes may be used, as with any airplane, to slow down, stop, or turn shorter than normal while taxiing. When initiating a turn though, they should be used cautiously to prevent overcontrolling of the turn. No airplane should be pivoted on one wheel when making sharp turns, as this can damage the landing gear, tires, or the airport pavement. All turns should be made with the inside wheel rolling, even if only slightly.

Brakes should be used as lightly as practicable while taxiing to prevent undue wear and heating of the brake discs. When brakes are used repeatedly or constantly they tend to heat to the point that they may either lock or fail completely. Also, tires may be weakened or blown out by extremely hot brakes. And any abrupt use of brakes in multiengine as well as single-engine airplanes, is evidence of poor pilot technique. The FAA frowns on this during checkrides.

Due to the greater weight of multiengine airplanes, effective braking is particularly essential. Therefore, as the airplane begins to move forward when taxiing is started, the brakes should be tested immediately by depressing each brake pedal. If you find the brakes are weak, it may be wise to have them checked.

Looking outside the cockpit while taxiing becomes even more important in multiengine airplanes. Since these airplanes are usually somewhat heavier, larger, and more powerful than single-engine airplanes they often require more time and distance to accelerate or stop, and provide a different perspective for the pilot. While it usually is not necessary to make S-turns to observe the taxiing path, additional vigilance is necessary to avoid obstacles, other aircraft, or bystanders.

Trim Tabs

The trim tabs in a multiengine airplane serve the same purpose as in a single-engine airplane, but their function is usually more important to safe and efficient flight. In addition to the elevator trim tab, multiengine aircraft always have a very effective rudder trim tab to help alleviate the asymmetrical thrust produced when being flown with only one engine running. This is because of the greater control forces required to handle the asymmetrical thrust with one engine inoperative.

In some multiengine airplanes it taxes the pilot's strength to overpower an improperly set trim tab on takeoff, go-around, or with one engine inoperative. Many fatal accidents have occurred when pilots took off or attempted a go-around with the airplane trimmed for something other than normal flight. Therefore, prompt retrimming of the elevator and rudder trim tabs in the event of an emergency go-around from a landing approach, especially from a single-engine approach, is essential to the success of the flight.

Multiengine airplanes, like all airplanes, should be retrimmed in flight for each change of attitude, airspeed, power setting, or single-engine configuration. Without changes in trim, the airspeed will suffer and the flight profile will often be somewhat like a roller coaster.

PERFORMANCE CHARACTERISTICS

Normal Takeoff

There is virtually no difference between a takeoff in a multiengine airplane and one in a single-engine airplane. The controls of each class of airplane are operated the same; the multiple throttles of the multiengine airplane normally are treated as one compact power control and can usually be operated simultaneously with one hand.

It is very important that the pilot have a plan of action to cope with engine failure during takeoff. It is equally important that just prior to takeoff the pilot mentally review takeoff procedures, especially procedures pertaining to losing an engine. This mental review should consist of the engine-out minimum control speed (V_{mc}), the best all-engine rate of climb speed (V_y), the best single-engine rate of climb speed (V_{yse}), and what procedures will be followed if an engine fails prior to reaching minimum control speed. This first speed (V_{mc}) is the minimum airspeed at which safe directional control can be maintained with one engine inoperative and one engine operating at full power.

The multiengine pilot's primary concern on all takeoffs is the attainment of at least the engine-out minimum control speed prior to liftoff. Until this speed is achieved, directional control of the airplane in flight will be impossible after the failure of an engine, unless power is reduced immediately on the operating engine. If an engine fails before the engine-out minimum control speed is attained, the pilot has no choice but to close both throttles, abandon the takeoff, and direct complete attention to bringing the airplane to a safe stop on the ground.

The multiengine pilot's second concern on takeoff is the attainment in the least amount of time of the single-engine best rate-of-climb speed (V_{yse}). This is the airspeed which will provide the greatest rate of climb when operating with one engine out and feathered (if possible), or the slowest rate of descent. In the event of an engine failure, the single-engine best rate-of-climb speed must be held until a safe maneuvering altitude is reached, or until a landing approach is initiated. The V_{yse} speed is shown on the airspeed indicator as a blue line across the indicator.

The engine-out minimum control speed (V_{mc}) and the single-engine best rate-of-climb speed (V_{yse}) are published in the airplane's FAA-approved flight manual. These speeds must be addressed mentally by the pilot before every takeoff (See FIG. 9-3).

Crosswind Takeoffs

Crosswind takeoffs are performed in multiengine airplanes in basically the same manner as those in single-engine airplanes. At the beginning of the takeoff roll, less power can be used on the downwind engine to overcome the tendency of the airplane to weathervane into the wind, and then full power applied to both engines as the airplane accelerates to a speed where better rudder control is attained.

Stalls

As with a single-engine airplane, the pilot should be familiar with the stall and minimum controllability characteristics of the multiengine airplane being flown. The larger and

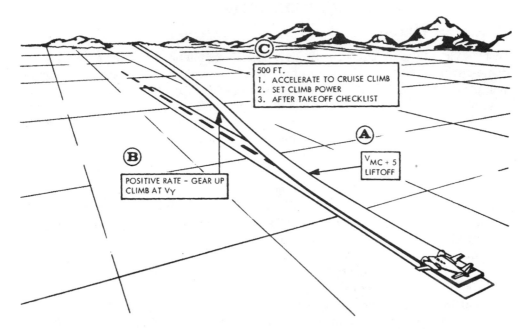

Fig. 9-3. *A multiengine takeoff profile requires reaching V_{mc}+5 knots prior to liftoff.*

heavier airplanes have slower responses in stalls and recoveries and in maneuvering at critically slow speeds due to their greater weight. The practice of stalls in multiengine airplanes, therefore, should be performed at altitudes sufficiently high to allow recoveries to be completed at least 3,000 feet above the ground.

It usually is inadvisable to execute full stalls in multiengine airplanes because of their relatively high wing loading. Therefore, practice should be limited to approaches to stalls (imminent), with recoveries initiated at the first physical indication of the stall. As a general rule, however, full stalls in multiengine airplanes are not necessarily violent or hazardous.

The pilot should become familiar with imminent stalls entered with various flap settings, power settings, and landing gear positions. It should be noted that the extension of the landing gear will cause little difference in the stalling speed, but it will cause a more rapid loss of speed in a stall approach.

Power-on stalls should be entered with both engines set at approximately 65 percent power. Takeoff power may be used provided the entry speed is not greater than the normal lift-off speed. Stalls in airplanes with relative low power loading using maximum climb power usually result in an excessive nose-high attitude and make the recovery more difficult. Additionally, some t-tail type multiengine aircraft tend to blank out their tails if the nose attitude is too high.

Because of possible loss of control, stalls with one engine inoperative or at idle power and the other developing effective power are not to be performed during multiengine flight tests nor should they be practiced by applicants for multiengine ratings. This is a dangerous practice and should be avoided at all times.

The same techniques used in recognition and avoidance of stalls of single-engine airplanes apply to stalls in multiengine airplanes. The pilot must be familiar with the characteristics which announce an approaching or imminent stall, the signals which the aircraft sends, and the proper technique for a positive recovery.

As with all aircraft, the increase in pitch attitude for stall entries should be gradual to prevent the airplane from climbing at an abnormally high nose-up attitude at the time the stall occurs. It is recommended that the rate of pitch change result in a one knot-per-second decrease in airspeed. In all stall recoveries the controls should be used very smoothly, avoiding abrupt pitch changes.

Slow Flight

Smooth control manipulation is important in all flight at minimum or critically slow airspeeds. As with all piloting operations, a smooth technique permits the development of a more sensitive feel of the controls with a keener sense of stall anticipation. Flight at minimum or critically slow airspeeds gives the pilot an understanding of the relationship between the decreasing control effectiveness and airspeed. The slower the aircraft travels through the air, the less effective will be the controls

Generally, the technique of flight at minimum airspeeds is the same in a multiengine airplane as it is in a single-engine airplane. Because of the additional equipment in the multiengine airplane, the transitioning pilot has more to do and observe, and the usually slower control reaction requires better anticipation. Care must be taken to observe engine temperature indications for possible overheating, and to make necessary power adjustments smoothly on both engines at the same time.

Approaches and Landings

Multiengine aircraft characteristically have steeper gliding angles than single-engine aircraft because of their relatively high wing loading and greater drag with wing flaps and landing gear when extended. For this reason, power is used throughout the approach to shallow the approach angle and prevent a high sink rate.

The accepted technique for making a stabilized landing approach is to reduce the power to a predetermined setting during the arrival descent so the appropriate landing gear extension speed (V_{lo}) will be attained in level flight as the downwind leg of the approach pattern is entered. With this power setting, the extension of the landing gear (when the airplane is on the downwind leg opposite the intended point of touchdown) will further reduce the airspeed to the desired traffic pattern airspeed. When within the maximum speed for flap extension (V_{fe}), the flaps may be partially lowered if desired, to aid in reducing the airspeed to traffic pattern speed.

The prelanding checklist should be completed by the time the airplane is on base leg so that the pilot may direct full attention to the approach and landing. In a powered approach, the airplane should descend at a stabilized rate, allowing the pilot to plan and control the approach path to the point of touchdown. Further extension of the flaps and slight adjustment of power and pitch should be accomplished as necessary to establish and maintain a stabilized approach path. The stabilized, powered approach should allow for the altitude to be controlled with power adjustments and the airspeed to be controlled with small pitch inputs.

The airspeed of the final approach should be the airspeed that provides the engine-out best rate-of-climb speed (V_{yse}) until the landing is assured, because that is the minimum speed at which a single-engine go-around can be made if necessary. In no case should the approach speed be less than the critical engine-out minimum control airspeed (V_{mc}). If an engine should fail suddenly and it is necessary to make a go-around from a final approach at less than that speed, a catastrophic loss of control would occur. As a rule of thumb, after the wing flaps are extended the final approach speed should be gradually reduced to V_{mc}+5 knots or the airspeed recommended by the manufacturer.

The roundout or flare should be started at sufficient altitude to allow a smooth transition from the approach to the landing attitude. The touchdown should be smooth, with the airplane touching down on the main wheels and the airplane in a tail-low attitude as the power is reduced to idle.

Directional control on the rollout should be accomplished primarily with the rudder and the steerable nosewheel, with discrete use of the brakes applied only as necessary for crosswinds or other factors.

Crosswind Landings

The crosswind landing technique in multiengine airplanes is little different from that required in single-engine airplanes. The only significant difference lies in the fact that because of the greater weight, more positive drift correction must be maintained before the touchdown.

It should be remembered that the FAA requires that most airplanes have satisfactory control capabilities when landing in a direct crosswind of at least 20 percent of the landing configuration stall speed ($0.2\ V_{so}$). Thus, an airplane with a power-off stalling speed of 60 knots has been designed to handle at least a direct crosswind of 12 knots ($.2 \times 60$) on landings. Though skillful pilots may successfully land in much stronger crosswinds, poor pilot technique may cause serious damage in less wind.

The two basic methods of making crosswind landings, the slipping approach (wing-low), and the crabbing approach may be combined. The essential factor in all crosswind-landing procedures is touching down without drift, with the heading of the airplane parallel to its direction of motion. This will result in minimum side loads on the landing gear.

Go-Around Procedure

The complexity of modern multiengine airplanes makes a knowledge of and proficiency in emergency go-around procedures particularly essential for safe piloting. The emergency go-around during a landing approach is inherently critical because it is usually initiated at a very low altitude and airspeed with the airplane configured for landing.

Unless absolutely necessary, the decision to go around should not be delayed to the point where the airplane is ready to touch down. The more altitude and time available to apply power, establish a climb, retrim, and set up a go-around configuration, the easier and safer the maneuver becomes. When the pilot has decided to go around, immediate action should be taken without hesitation, while maintaining positive control and accurately following the manufacturer's recommended procedures.

Go-around procedures can vary with different airplanes, depending on their weight, flight characteristics, flap and retractable gear systems, and flight performance. And there are several go-around procedures which can apply to most multiengine airplanes. Although there may be slight differences, most twins follow the same basic procedures. When the decision to go around is reached, takeoff power should be applied immediately and the descent stopped by adjusting the pitch attitude to avoid further loss of attitude.

The flaps should be retracted only in accordance with the procedure prescribed in the airplane's operating manual. Usually this will require the flaps to be raised to near take-off position.

After a positive rate of climb is established the landing gear should be retracted, best single-engine rate-of-climb airspeed obtained and maintained, and the airplane trimmed for this climb. The procedure for a normal takeoff climb should then be followed.

The basic requirements of a successful go-around, then, are to power up, pitch up to arrest the descent, then attain and maintain the best rate-of-climb airspeed, clean the aircraft up, and proceed.

At critically slow airspeeds, retracting the flaps prematurely can cause an unanticipated loss of altitude. Rapid or premature retraction of the flaps should be avoided on go-arounds, especially when close to the ground, because of the careful attention and exercise of precise pilot technique necessary to prevent a sudden loss of altitude. It generally will be found that retracting the flaps only halfway or to the specified approach setting decreases the drag a relatively greater amount than it decreases the lift.

The FAA approved Airplane Flight Manual or Pilot's Operating Handbook should be consulted regarding landing gear and flap retraction procedures because in some installations simultaneous retraction of the gear and flaps may increase the flap retraction time, and full flaps create more drag than the extended landing gear.

MULTIENGINE ENGINE-OUT PROCEDURES

From the would-be multiengine pilot's point of view, the primary difference between a light-twin and a single-engine airplane is the potential problem involving an engine failure. The information that follows is designed to highlight that one potentially deadly issue of what happens when one engine fails.

V Speeds

Before operating techniques in light twin-engine airplanes can be fully discussed, the subject of "V" speeds and their relevance to multiengine flight must be addressed. "V" speeds such as V_{mc}, V_x, V_{xse}, V_y, and V_{yse} are the main performance speeds the light-twin pilot needs to know in addition to the other performance speeds common to both twin-engine and single-engine airplanes. The airspeed indicator in twin-engine airplanes is marked (in addition to other normally marked speeds) with a red radial line at the minimum controllable airspeed with the critical engine inoperative (V_{mc}), and a blue radial line at (V_{yse}),the best single-engine rate-of-climb airspeed with one engine inoperative (See FIG. 9-4).

V_x is the speed for best angle of climb. At this speed the airplane will gain the greatest amount of altitude for a given distance of forward travel. This speed is used for obstacle clearance with all engines operating. However, this speed is very different when one engine is inoperative, and is referred to as V_{xse}, or best angle of climb airspeed (single-engine).

V_y is the speed for the best rate of climb. This speed will provide the maximum altitude gain for a given period of time with all engines operating. However, this speed, too, will be different when one engine is inoperative and is referred to as V_{yse}, or best rate of climb (single-engine).

V_{mc}

V_{mc} is the minimum control speed with the critical engine inoperative. The term V_{mc} can be defined as the minimum airspeed at which the airplane is controllable when the critical engine is made suddenly inoperative, and the remaining engine is producing takeoff power. The Federal Aviation Regulations under which the airplane was certificated, stipulate that at V_{mc} the aircraft must be able to:

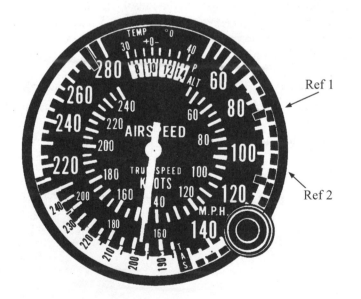

Fig. 9-4. *A typical twin-engine airspeed indicator contains two extra colored radials, the one at V_{mc} is red and the one at V_{yse} is blue.*

- Stop the turn, which results when the critical engine is suddenly made inoperative within 20 degrees of the original heading, using maximum rudder deflection and a maximum of 5-degrees bank into the operative engine.

- After recovery, maintain straight flight with not more than a 5-degree bank (wing lowered toward the operating engine).

This does not mean that the airplane must be able to climb or even hold altitude. It only means that a heading can be maintained. The principle of V_{mc} is not at all mysterious. It is simply that at any airspeed less than V_{mc}, air flowing along the rudder is such that application of rudder forces cannot overcome the asymmetrical yawing forces caused by takeoff power on one engine and a powerless windmilling propeller on the other.

Many pilots erroneously believe that because a light-twin has two engines, it will continue to perform at least half as well with only one of those engines operating. Wrong! There is nothing in FAR, Part 23, governing the certification of light-twins, which requires an airplane to maintain altitude while in the takeoff configuration and with one engine inoperative. In fact, many of the current light-twins are not required to do this with one engine inoperative in any configuration, even at sea level.

Single-Engine Performance

When one engine fails on a light-twin, performance is not only halved, it is actually reduced by 80 percent or more. The performance loss is greater than 50 percent because an airplane's climb performance is a function of the thrust horsepower which is in excess of that required for level flight. When power is increased in both engines in level flight and the airspeed is held constant, the airplane will start climbing. The rate of climb will depend on the power added (which is power in excess of that required for straight-and-level flight). When one engine fails, however, it not only loses power but the drag increases considerably because of asymmetric thrust and the operating engine must then carry the full burden alone. To do this, it must produce 75 percent or more of its rated power. This leaves very little excess power for climb performance, especially in a hot, humid weather condition or at high altitude.

As an example, an airplane which has an all-engine rate of climb of 1,860 FPM and a single-engine rate of climb of 190 FPM would lose almost 90 percent of its climb performance when one engine fails. Quite a difference!

Nonetheless, the light-twin does offer obvious safety advantages over the single-engine airplane (especially in the enroute phase) but only if the pilot fully understands the real options offered by that second engine in the takeoff and approach phase of flight.

Engine-Out Emergencies

In general, the operating and flight characteristics of modern light-twins with one engine inoperative are excellent. These airplanes can be controlled and maneuvered safely as long as sufficient airspeed is maintained. Note the key word is airspeed. Airspeed is life to a multiengine pilot operating on only one engine. However, to utilize the safety and performance characteristics effectively, the pilot must have a sound understanding of single-engine performance and limitations resulting from unequal thrust.

A pilot upgrading to a multiengine airplane should practice and become thoroughly familiar with the control and performance problems which result from the failure of one engine during any flight condition. This is really the major issue of multiengine flight. Proficiency in all the control operations and precautions must be demonstrated on multiengine rating flight tests.

The feathering of a propeller should be demonstrated and practiced in all airplanes equipped with propellers which can be feathered and unfeathered safely in flight. If the airplane used is not equipped with feathering propellers, or is equipped with propellers which cannot be feathered and unfeathered safely in flight, one engine should be secured (shut down) in accordance with the procedures in the FAA approved Airplane Flight Manual or the Pilot's Operating Handbook.

Propeller Feathering

When an engine fails in flight the movement of the airplane through the air tends to keep the propeller rotating, much like a windmill. Since the failed engine is no longer delivering power to the propeller to produce thrust, but instead, is absorbing energy to overcome friction and compression of the engine, the drag of the windmilling propeller is significant and causes the airplane to yaw toward the failed engine. Most multiengine airplanes are equipped with "full-feathering propellers" to minimize that yawing tendency.

The blades of a feathering propeller may be positioned by the pilot to such a high angle that they are streamlined in the direction of flight. In this feathered position, the blades act as powerful brakes to assist engine friction and compression in stopping the windmilling rotation of the propeller. This is of particular advantage in case of a damaged engine, since further damage, caused by a windmilling propeller creates the least possible drag on the airplane and reduces the yawing tendency. As a result, multiengine airplanes are easier to control in flight when the propeller of an inoperative engine is feathered.

Feathering of propellers should be performed only under such conditions and at such altitudes and locations that a safe landing on an established airport could be accomplished readily in the event of difficulty in unfeathering the propeller. I never feather a propeller below 3,000 feet above the ground or when I am not near (or over) an airport.

Engine-Out Procedures

The following procedures are recommended to develop in the multiengine pilot the habit of using proper procedures and proficiency in coping with an inoperative engine.

At a safe altitude (minimum 3,000 feet above terrain) and within gliding distance of a suitable airport, an engine may be shut down with the mixture control or fuel selector. At lower altitudes, however, engine shutdown should be simulated by reducing power by means of the throttle to the zero thrust setting. The following procedures should then be followed.

- Set the mixture and propeller controls full forward. Then both throttles should be increased to full throttle for maximum power to maintain at least V_{mc}.
- Retract wing flaps and landing gear to minimize drag.
- Determine which engine is failed, and then verify it by closing the throttle to the dead engine.
- Bank at least 5 degrees into the operative engine.
- Determine the cause of failure, or feather the inoperative engine.
- Turn toward the nearest airport.
- Secure (shut down) the inoperative engine in accordance with the manufacturer's checklist and check for engine fire.
- Monitor the engine instruments on the operating engine; and adjust power, cowl flaps, and airspeed as necessary.
- Maintain altitude and an airspeed of at least V_{yse} if possible.

The pilot must be proficient in the control of heading, airspeed, and altitude, in the prompt identification of a power failure, and in accuracy of shutdown and restart procedures as prescribed in the FAA approved Airplane Flight Manual.

There is no better way to develop skill in single-engine emergencies than by continued practice. The techniques and procedures of single-engine operation do not necessarily remain at a consistently high level. Like any other skill, lack of review and practice tends to erode fundamental procedures and skills may be lost. And some engine-out emergencies can be so critical that there may be no safety margin for lack of skill or knowledge. Unfortunately, many light-twin pilots never practice single-engine operation after receiving their multiengine rating.

The pilot should practice and demonstrate the effects of engine-out performance at various configurations of gear, flaps, and both; the use of carburetor heat; and the failure to feather the propeller on an inoperative engine. Each configuration should be maintained, at best engine-out rate-of-climb speed long enough to determine its effect on the climb (or sink) achieved.

THE CRITICAL ENGINE

"P-factor" is present in multiengine airplanes just as it is in single-engine airplanes. P-factor is caused by the unequal thrust of rotating propeller blades when the airplane is in a nose-high, power-on configuration. It is the result of the descending blade having a greater angle of attack than the ascending blade when the relative wind striking the blades is not aligned with the thrust line (as in a nose-high attitude).

In many U.S. designed light-twins, both engines rotate to the right (clockwise) when viewed from the cockpit, and both engines develop an equal amount of thrust. At high angles of attack and high-power conditions, the downward moving propeller blade of both engines develop more thrust than the upward moving blade. This asymmetric propeller thrust or "P-factor," results in a center of thrust being at the right side on both engines

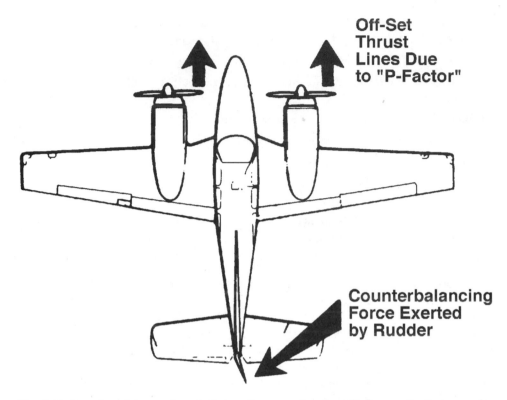

Off-Set Thrust Lines Due to "P-Factor"

Counterbalancing Force Exerted by Rudder

Fig. 9-5. *On twins whose engines both turn the same direction, the forces of p-factor produce a more pronounced left-turning tendency since the force is double that of a single-engine aircraft.*

(See FIG. 9-5). The turning (or yawing) force of the right engine is greater than the left engine because the center of thrust is much farther away from the center line of the fuselage giving it a longer arm.

Thus, when the right engine is running and the left engine is inoperative, the turning (or yawing) force is greater than in the opposite situation of an operating left engine and an inoperative right engine. In other words, directional control may be difficult when the left engine (the critical engine) is suddenly made inoperative.

It should be noted that many newer twin-engine airplanes are equipped with engines turning in opposite directions; that is, the left engine and propeller turn clockwise and the right engine and propeller turn counterclockwise. With this arrangement, the thrust line of either engine is the same distance from the center line of the fuselage, so there will be no difference in yaw effect between loss of left or right engine. This effectively eliminates this type of airplane having a critical engine and it makes them much safer when an engine is lost.

V_{mc} Demonstrations

Multiengine instruction must include a demonstration of the airplane's engine-out minimum control speed (V_{mc}). The engine-out minimum control speed given in the Airplane

Flight Manual is determined during original airplane certification under conditions specified by the FAA. These conditions normally are not duplicated during pilot training because they consist of the most adverse situations for airplane type certification purposes.

Basically, when one engine fails the pilot must overcome the asymmetrical thrust created by the operating engine by setting up a counteracting moment with the rudder. When the rudder is fully deflected, its yawing power will depend on the velocity of airflow across the rudder, which in turn is dependent on airspeed. As the airplane decelerates it will reach a speed below which the rudder command will no longer balance the unequal thrust and directional control will be lost. This is V_{mc}.

V_{mc} is shown on the airspeed indicator as a red radial line. The indicated V_{mc} speed is as high as it will ever get. It can be lower, but never higher. V_{mc} is determined with the aircraft in a configuration which will create the worst possible conditions relative to the sudden loss of one engine. These conditions are:

- Takeoff power on both engines.
- Most rearward center of gravity.
- Flaps in the takeoff setting.
- Landing gear retracted.
- Propeller windmilling.

The V_{mc} demonstrations should be performed at an altitude of at least 3,000 feet above the surface, and preferably near an airport. One demonstration should be made while holding the wings level and the ball centered, and another demonstration should be made while banking the airplane at least 5 degrees toward the operating engine to establish "zero sideslip." These maneuvers will demonstrate the engine-out minimum control speed for the existing conditions and will emphasize the necessity of banking into the operative engine. No attempt should be made to duplicate V_{mc} as determined for airplane certification.

After the propellers are set to high RPM, the landing gear is retracted, and the flaps are in the takeoff position, the airplane should be placed in a climb attitude and airspeed at V_{yse}. With both engines developing takeoff power, power on the critical engine (usually the left) should then be reduced to idle. After this is accomplished, the airspeed should be reduced about one knot per second with the elevators until directional control no longer can be maintained.

At this point, recovery should be initiated by simultaneously reducing power on the operating engine and reducing the angle of attack by lowering the nose. Allow the airspeed to build back just above V_{mc} and then increase the power on the good engine as the aircraft is banked about five degrees toward the good engine. Slowly raise the nose to level flight attitude, maintain V_{yse}, and set the rudder trim tab to aid in directional control.

Should indications of a stall occur prior to reaching this point, recovery should be initiated immediately by reducing the angle of attack. In this case, a minimum engine-out control speed demonstration is not possible under existing conditions.

If it is found that the minimum engine-out control speed is reached before indications of a stall are encountered, the pilot should demonstrate the ability to control the airplane and initiate a safe climb in the event of a power failure at the published engine-out minimum control speed.

Lateral Lift

During engine-out flight the large rudder deflection required to counteract the asymmetrical thrust also results in a "lateral lift" force on the vertical fin. This lateral "lift" represents an unbalanced side force on the airplane which must be countered either by allowing the airplane to accelerate sideways until the lateral drag caused by the sideslip equals the rudder "lift" force or by banking into the operative engine and using a component of the airplane weight to counteract the rudder-induced side force.

In the first case, the wings will be level, the ball in the turn-and-slip indicator will be centered and the airplane will be in a moderate sideslip toward the inoperative engine. In the second case, the wings will be banked 3–5 degrees into the good engine, the ball will be deflected one diameter toward the operative engine, and the airplane will be at zero sideslip.

The sideslipping method has several major disadvantages:

- The relative wind blowing on the inoperative engine side of the vertical fin tends to increase the asymmetric moment caused by the failure of one engine.

- The resulting sideslip severely degrades stall characteristics.

- The greater rudder deflection required to balance the extra moment and the sideslip drag cause a significant reduction in climb and/or acceleration capability.

Flight tests have shown that holding the ball of the turn-and-slip indicator in the center while maintaining heading with wings level drastically increases V_{mc} as much as 20 knots in some airplanes. (Remember, the value of V_{mc} given in the FAA approved flight manual for the airplane is based on a maximum 5 degrees bank into the operative engine.) Banking into the operative engine reduces V_{mc}, whereas decreasing the bank angle away from the operative engine increases V_{mc} at the rate of approximately 3 knots per degree of bank angle.

Flight tests have also shown that the high drag caused by the wings level, ball-centered configuration can reduce single-engine climb performance by as much as 250 FPM, which may be more than is available at sea level in a nonturbocharged light twin.

Banking at least 5 degrees into the good engine ensures that the airplane will be controllable at any speed above the certificated V_{mc}, that the airplane will be in a minimum drag configuration for best climb performance, and that the stall characteristics will not be degraded. Engine-out flight with the ball centered is never correct.

For an airplane with nonsupercharged engines, V_{mc} decreases as altitude is increased. Consequently, directional control can be maintained at a lower airspeed than at sea level. The reason for this is that since power decreases with altitude the thrust moment of the operating engine becomes less, thereby lessening the need for the rudder's yawing force. Since V_{mc} is a function of power (which decreases with altitude), it is possible for the airplane to reach a stall speed prior to loss of directional control.

Stall Prior To V$_{mc}$

It must be understood that there is a certain density altitude above which the stalling speed is higher than the engine-out minimum control speed. When this density altitude exists close to the ground because of high elevations or temperatures, an effective flight demonstration of V$_{mc}$ is impossible and should not be attempted. When a flight demonstration is impossible, the instructor should emphasize orally the significance of the engine-out minimum control speed, including the results of attempting flight below this speed with one engine inoperative, the recognition of the imminent loss of control, and the recovery techniques involved.

Relationship of V$_{mc}$ to CG

V$_{mc}$ is greater when the center of gravity is at the rearmost allowable position. Since the airplane rotates around its center of gravity, the moments are measured using that point as a reference. A rearward CG would not affect the thrust moment, but would shorten the arm to the center of the rudder's horizontal "lift" which would mean that a higher force (airspeed) would be required to counteract the engine-out yaw.

Generally, the center of gravity range of most light twins is short enough so that the effect on the V$_{mc}$ is relatively small, but it is a factor that should be considered. Many pilots would only consider the rear CG of their light-twin as a factor for pitch stability, not realizing that it could greatly affect the controllability with one engine out.

There are many light-twin pilots who think that the only control problem experienced in flight below V$_{mc}$ is a yaw toward the inoperative engine. Unfortunately, this is not the whole story.

With full power applied to the operative engine, as the airspeed drops below V$_{mc}$, the airplane tends to roll as well as yaw into the inoperative engine. This tendency becomes greater as the airspeed is further reduced. Since this tendency must be counteracted by aileron control, the yaw condition is aggravated by adverse yaw (the "down" aileron creates more drag than the "up" aileron). If a stall should occur in this condition, a violent roll into the dead engine may be experienced. Such an event occurring close to the ground would be disastrous. This can be avoided by keeping the airspeed above V$_{mc}$ at all times during single-engine operation. If the airspeed should fall below V$_{mc}$ for whatever reason, power must be reduced on the operative engine, and the nose must be lowered to regain airspeed above V$_{mc}$ (See FIG. 9-6). At that point, power may be reapplied to the operating engine and the airplane can be carefully controlled.

ZONE OF DECISION

The most critical time for an engine-out condition in a twin-engine airplane is during the few seconds immediately following liftoff while the airplane is accelerating to a safe engine-failure speed.

This "zone of decision" is bounded by the point at which V$_y$ is reached and the point where the obstruction altitude is reached. An engine failure in this area demands

Fig. 9-6. *The rudder trim saves a lot of legwork when flying on a single engine.*

an immediate decision to abort the flight, land and stop, or to continue the flight. Any indecision in this zone will probably end tragically. Beyond this decision area, the pilot has but one choice, to continue with the takeoff, maneuver carefully to whatever altitude can be attained, and work carefully through the process of returning to the airport. Here is where patience is a true virtue. Don't attempt to force performance from the aircraft or get in a hurry to get back. If the aircraft is flying and has no obstacles in front of it, altitude can usually be coaxed from the aircraft.

Although most twin-engine airplanes are controllable at a speed close to the engine-out minimum control speed, the performance is so far below optimum that continued flight following takeoff may be marginal or impossible. A more suitable recommended speed, termed by some aircraft manufacturers as minimum safe single-engine speed (V_{sse}), is that speed at which altitude can be maintained while the landing gear is being retracted and the propeller is being feathered.

Upon engine failure after reaching the safe single-engine speed on takeoff, the twin-engine pilot (having lost one-half of the normal power) usually has a significant advantage over the pilot of a single-engine airplane. This is because, if the aircraft has single-engine climb capability at the existing gross weight and density altitude, there may be the choice of stopping or continuing the takeoff. This compares with the only choice facing a single-engine airplane pilot who suddenly has lost half of the normal takeoff power. They have to stop!

Factors in a Go No-Go Decision

If one engine fails prior to reaching V_{mc}, the decision has already been made to close both throttles and bring the airplane to a stop on the runway. If engine failure occurs after becoming airborne, the pilot must decide immediately whether to land or to continue the flight.

If the decision is made to continue the flight, the airplane must be able to at least hold its altitude with one engine inoperative. This requires acceleration to V_{yse} if no obstacles are involved, or to V_{xse} if obstacles are a factor.

To make a correct decision in an emergency of this type, the pilot should have considered the runway length, field elevation, density altitude, obstruction height, headwind component, and the airplane's gross weight prior to takeoff. This would not be an enviable position to be in without a prior plan.

Accelerate/Stop Distance

The "accelerate-stop distance" is the total distance required to accelerate the twin-engine airplane to liftoff speed and, assuming failure of an engine at the instant that speed is attained, bring the airplane to a stop on the remaining runway.

The "accelerate-go distance" is the total distance required to accelerate the airplane to a specified speed and, assuming failure of an engine at the instant that speed is attained, continue takeoff on the remaining engine to a height of 50 feet.

For example, assume that with a temperature of 80 degrees F., a calm wind at a pressure altitude of 2,000 feet, a gross weight of 4,800 pounds, and all engines operating, the airplane being flown requires 3,525 feet to accelerate to 105 MPH and then be brought to a stop. Let's also assume that the airplane under the same conditions requires a distance of 3,830 feet to take off and climb over a 50-foot obstacle when one engine fails at 105 MPH.

With such a slight margin of safety (305 feet) it would be better to discontinue the takeoff and stop if the runway is of adequate length, since any slight mismanagement of the engine-out procedure would more than outweigh the small advantage offered by continuing the takeoff. At higher field elevations the advantage becomes less and less until at very high density altitudes a successful continuation of the takeoff is extremely improbable.

OTHER FACTORS IN TAKEOFF PLANNING

Proficient pilots of light-twins plan the takeoff in sufficient detail to be able to take immediate action if one engine fails during the takeoff process. They are thoroughly familiar with the airplane's performance capabilities and limitations, including accelerate-stop distance, as well as the distance available for takeoff, and will include such factors in their plan of action. For example, if it has been determined that the airplane cannot maintain altitude with one engine operative (considering the gross weight and density altitude), the seasoned pilot will be well aware that should an engine fail right after lift-off,

Fig. 9-7. *A thorough preflight will do more to prevent a multiengine flight from becoming a single-engine flight than anything I can think of.*

an immediate landing will have to be made in the most suitable area available. The sane pilot will not attempt to maintain altitude at the expense of a safe airspeed. This amounts to suicide.

Consideration will also be given to surrounding terrain, obstructions, and nearby landing areas so that a definite direction of flight can be established immediately if an engine fails at a critical point during the climb after takeoff. It is imperative then, that the takeoff and climb path be planned so that all obstacles between the point of takeoff and the available areas of landing can be cleared if one engine suddenly becomes inoperative.

Airspeed Control

The twin-engine airplane must be flown at precise airspeeds if maximum takeoff performance and safety are to be obtained. For example, the airplane must lift off at its specific liftoff airspeed, accelerate to V_y airspeed, and climb with maximum permissible power on both engines to a safe single-engine maneuvering altitude. Prior to that, if an engine fails, a different airspeed must be attained immediately, usually V_{yse}. This airspeed must be held precisely because only at this airspeed will the pilot be able to obtain maximum performance from the airplane. To understand the factors involved in proper takeoff planning, a further explanation of this critical speed follows, beginning with the lift-off.

I have said several times the light-twin can be controlled satisfactorily while firmly on the ground when one engine fails prior to reaching V_{mc} during the takeoff roll. This is possible by closing both throttles, by proper use of rudder, brakes, and nosewheel steering. If the airplane should be airborne at less than V_{mc}, however, and suddenly loses all power on one engine, it cannot be controlled satisfactorily. Thus, on normal takeoffs, lift-off should never take place until the airspeed reaches and exceeds V_{mc}. The FAA recommends a minimum speed of V_{mc} plus 5 knots before lift-off.

An efficient climb procedure is one in which the airplane leaves the ground at V_{mc}+5 knots, accelerates quickly to V_{yse} (best rate-of-climb speed, single-engine), and then accelerates to V_y. The climb at V_y should be made with both engines set to maximum take-off power until reaching a safe single-engine maneuvering altitude, a minimum of 500' above field elevation. At this point, power may be reduced to the allowable maximum continuous power setting (METO-maximum except takeoff) or less, and any desired en-route climb speed then may be established.

Airspeed vs. Altitude

In the event of engine failure, a pilot who uses excessive speed on takeoff will discover suddenly that all the energy produced by the engines has been converted into speed. Although airspeed is important, excessive airspeed is not nearly so important as is altitude. Especially below 500 feet, or so.

Improperly trained pilots often believe that the excess speed can always be converted to altitude, but this theory is not correct. Available power is always wasted in accelerating the airplane to an unnecessary speed. Also, experience has shown that an unexpected engine failure so surprises the unseasoned pilot that proper reactions are often extremely slow in coming. By the time the initial shock wears off and the pilot is ready to take control of the situation, the excess speed has dissipated and the airplane is still barely off the ground. From this low altitude, the pilot still has to climb, with an engine inoperative, to whatever height is needed to clear all obstacles and get back to the approach end of the runway. The prospect for a successful conclusion from this mismanagement of airspeed is dim.

The laws of physics dictate that the airplane will expend less energy to fly in level flight than it will to climb. Therefore, if the total energy of both engines is initially converted to enough height above the ground to permit clearance of all obstacles (safe maneuvering altitude), the problem is much simpler in the event an engine fails. If some extra altitude is available, it can always be traded for airspeed or gliding distance when needed.

There is a fine line between too much or too little airspeed or altitude. Usually both are the pilot's friends, and one usually complements the other. But, in the time just after liftoff, until the aircraft reaches about 500 feet, or so, altitude is more important than airspeed. However, the pilot cannot place so much emphasis on altitude that airspeed strays far from the desired amount. There must be a balance between enough altitude and enough airspeed that is brought about by paying strict attention to all revealing information.

Consequently, during takeoff and early climb, the pilot should always be ready for any eventuality by keeping one hand on the control wheel and the other hand on the throttle. The airplane must remain on the ground until V_{mc} + 5 knots is reached so that a smooth transition to the proper climb speed can be made. THE AIRPLANE SHOULD NEVER LEAVE THE GROUND BEFORE V_{mc} IS REACHED.

If an engine fails before leaving the ground it is advisable to discontinue the takeoff and stop. If an engine fails after lift-off, the pilot will have to decide immediately whether to continue flight, or to close both throttles and land. However, waiting until the engine failure occurs is not the time for the pilot to plan the correct action. The action must be planned before the airplane is taxied onto the runway. The plan of action must consider the density altitude, length of the runway, weight of the airplane, and the airplane's accelerate-stop distance, and accelerate-go distance under these conditions. Only on the basis of these factors can the pilot decide intelligently what course to follow if an engine should fail.

To reach a safe single-engine maneuvering altitude as safely and quickly as possible, the climb with all engines operating must be made at the proper airspeed, usually V_y. That speed will provide for:

- Good control for the airplane in case an engine fails.
- Quick and easy transition to the single-engine best rate-of-climb speed if one engine fails.
- A fast rate of climb to attain an altitude which permits adequate time for analyzing the situation and making decisions.

To make a quick and easy transition to the single-engine best rate-of-climb speed, in case an engine fails, the pilot should climb at some speed greater than V_{yse}, probably V_y. If an engine fails at less than V_{yse}, it would be necessary for the pilot to lower the nose to increase the speed to V_{yse} in order to obtain the best climb performance. If the climb airspeed is considerably less than this speed, it might be necessary to lose valuable altitude to increase the speed to V_{yse}. Another factor to consider is the loss of airspeed that may occur because of erratic pilot technique after a sudden, unexpected power loss. Consequently, the normal initial two-engine climb speed should not be less than V_y.

NORMAL TAKEOFF PROCEDURES

After runup and pretakeoff checks have been completed, the airplane should be taxied into takeoff position and aligned with the runway. If the crew consists of two pilots, the pilot in command should brief the other pilot on takeoff procedures prior to receiving clearance for takeoff. This briefing consists of at least the following:

- Minimum controllable airspeed (V_{mc}).
- Rotation speed (V_r).
- Liftoff speed (V_{lof}).
- Single-engine best rate-of-climb speed (V_{yse}).
- All-engine best rate-of-climb speed (V_y).
- What procedures will be followed if an engine failure occurs prior to V_{mc}.

Both throttles then should be advanced simultaneously to takeoff power, and directional control maintained by the use of the steerable nosewheel and the rudder. Brakes should be used for directional control only during the initial portion of the takeoff roll when the rudder and steerable nosewheel are ineffective.

As the takeoff progresses, flight controls are used as necessary to compensate for wind conditions. Lift-off should be made at no less than V_{mc}+5 knots. After lift-off, the airplane should be allowed to accelerate to the single-engine best rate of climb speed, V_{yse}, and then accelerated to the all-engine best rate-of-climb speed, V_y, and then to climb maintaining this speed with takeoff power until a safe maneuvering altitude is attained.

The landing gear may be raised as soon as practicable, but not before reaching the point from which a safe landing can no longer be made on the remaining runway. The flaps should be retracted as directed in the airplane's operating manual.

Upon reaching a safe maneuvering altitude, the airplane should be allowed to accelerate to cruise climb speed before power is reduced for normal climb power, cowl flaps are adjusted, and trims tabs are reset.

Short Field or Obstacle Takeoff

When it is necessary to take off over an obstacle or from a critically short field, the procedures will be altered slightly from a normal takeoff. For example, the initial climb speed that will provide the best angle of climb for obstacle clearance is V_x rather than V_y. Additionally, every inch of runway should be utilized because runway behind you is never useful.

Generally, use the aircraft manufacturer's recommended airspeed, flap, and power settings. However, if the published best angle-of-climb speed (V_x) is less than V_{mc}+5, then the prudent pilot uses no less than V_{mc}+5 knots for liftoff and initial climb.

During the takeoff roll as the airspeed reaches the best angle-of-climb speed, or V_{mc}+5, whichever is higher, the airplane should be firmly rotated to establish an angle of attack that will cause the airplane to lift off and climb at that specified speed. At an altitude of approximately 50 feet, or after clearing the obstacle, the pitch attitude can be lowered gradually to allow the airspeed to increase to the all-engine best rate-of-climb speed. Upon reaching safe maneuvering altitude, the airplane should be allowed to accelerate to normal or enroute climb speed and the power reduced to the normal climb power settings, cowl flaps adjusted and aircraft retrimmed.

Engine Failure During Takeoff Roll

If an engine should fail during the takeoff roll before becoming airborne, the pilot should close both throttles immediately and bring the airplane to a stop. The same procedure is recommended if, after becoming airborne, an engine should fail prior to having reached the single-engine best rate-of-climb speed (V_{yse}). An immediate landing is inevitably safer because the altitude loss required to increase the speed to V_{yse} usually will preclude obstacle avoidance.

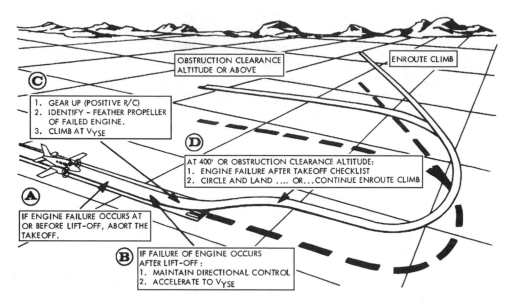

Fig. 9-8. *If an engine fails prior to attaining V_{mc}, abort the takeoff (A). An engine failure after reaching V_{mc} requires preplanning, proficiency, and patience.*

The pilot must have determined before takeoff what altitude, airspeed, and airplane configuration must exist to permit the flight to continue in event of an engine failure (See FIG. 9-8). The pilot also should be ready to accept the fact if engine failure occurs before these required factors are established, both throttles must be closed and the situation treated the same as engine failure on a single-engine airplane. That is, the pilot will have to land straight ahead and as slowly as possible to enhance the chances of surviving the forced landing.

If it has been predetermined that the engine-out rate of climb under existing circumstances will be at least 50 feet per minute at 1,000 feet above the airport, and that at least the engine-out best angle-of-climb speed has been attained, the pilot may decide to continue the takeoff.

. If the airspeed is below the engine-out best angle-of-climb speed (V_{xse}) and the landing gear has not been retracted, the takeoff should be abandoned immediately.

Engine Failure During Climbout

If the engine-out best angle-of-climb speed (V_{xse}) has been obtained and the landing gear is in the retract cycle, the pilot should climb at the engine-out best angle-of-climb speed (V_{xse}) to clear any obstructions, and thereafter stabilize the airspeed at the engine-out best rate-of-climb speed (V_{yse}) while retracting the landing gear and flaps and maintaining aircraft control. But at all costs, fly the airplane. Don't trade the job as pilot for one as passenger during an emergency, it is usually fatal.

If the decision is made to continue flight, the single-engine best rate-of-climb speed should be attained and maintained. Even if altitude cannot be maintained, it is best to continue to hold that speed because it would result in the slowest rate of descent and provide the most time for executing the emergency landing. After the decision is made to continue flight and a positive rate of climb is attained, the landing gear should be retracted as soon as practical.

If the airplane is barely maintaining altitude and airspeed, it is wise to avoid any attempt to turn the aircraft. When a turn is made under these conditions, both lift and airspeed decrease and an altitude loss is inevitable. Consequently, continue straight ahead whenever possible until reaching a safe maneuvering altitude. Pilots have died for lack of patience in this most gut-wrenching of times.

When an engine fails after becoming airborne, the pilot should hold heading with rudder and simultaneously roll into a bank of at least 5 degrees toward the operating engine. In this attitude the airplane will tend to turn toward the operating engine, but at the same time, the asymmetrical power resulting from the engine failure will tend to turn the airplane toward the "dead" engine. The result is a partial balance of those tendencies and provides for an increase in airplane performance as well as easier directional control.

The best way to identify the inoperative engine is to note the direction of yaw and the rudder pressure required to maintain heading. To counteract the asymmetrical thrust, extra rudder pressure will have to be exerted on the operating engine side. To aid in identifying the failed engine, some pilots use the expression "Idle Foot-Idle Engine," meaning the engine paired to the pilot's foot that is not pressing the rudder is the dead engine. Never rely on tachometer or manifold pressure readings to determine which engine has failed. After power has been lost on an engine, the manifold pressure will indicate the approximate atmospheric pressure which often looks remarkably the same as climb power. Don't be fooled by this.

Experience has shown that the biggest problem is not in identifying the inoperative engine, but rather in the pilot's actions after the inoperative engine has been identified. In other words, a pilot may identify the "dead" engine and then attempt to shut down the good one. This leads to a rather quiet airplane and a very rapid pulse.

To avoid this mistake, the pilot should verify that the dead engine has been identified by s-l-o-w-l-y retarding the throttle of the suspected engine before shutting it down and feathering the propeller. Must I explain why slowly is the proper speed for this action? Think about it.

When demonstrating or practicing procedures for engine failure on takeoff, the feathering of the propeller and securing of the engine should be simulated rather than actually performed, so that the engine may be available for immediate use if needed.

In all cases, the airplane manufacturer's recommended procedure for single-engine operation should be followed. The general procedure listed below is not intended to replace or conflict with any procedure established by the manufacturer of any airplane. It can be used effectively for general training purposes and to emphasize the importance of V_{yse}. It should be noted that this procedure is concerned with an engine failure on a takeoff where obstacle clearances are not critical. If the decision is made to continue flight after an engine failure during the takeoff climb, the pilot should maintain directional control at all times, maintain V_{yse}, and:

- Check that all mixture controls, prop controls, and throttles (in that order) are at maximum permissible power settings.
- Check that the flaps and landing gear have been retracted.
- Decide which engine is inoperative (dead).
- Raise the wing on the suspected "dead" engine at least 5 degrees.
- Verify the "dead" engine by retarding the throttle of the suspected engine. (If there is no change in rudder forces, then that is the inoperative engine.)
- Feather the prop on the "dead" engine (verified by the retarded throttle).
- Declare an emergency if operating from a tower-controlled airport. Advise the tower of your intentions.

ENGINE FAILURE ENROUTE

Normally, when an engine failure occurs while enroute at cruising flight altitude, the situation is not as critical as when an engine fails on takeoff. Having plenty of altitude and airspeed, the pilot can take time to determine the cause of the failure and remedy the condition, if possible. If the condition cannot be corrected, the single-engine procedure recommended by the manufacturer should be accomplished and a landing made as soon as practical (See FIG. 9-9).

A primary error during any engine failure is the pilot's tendency to perform the engine-out identification and shutdown too quickly, resulting in improper identification or incor-

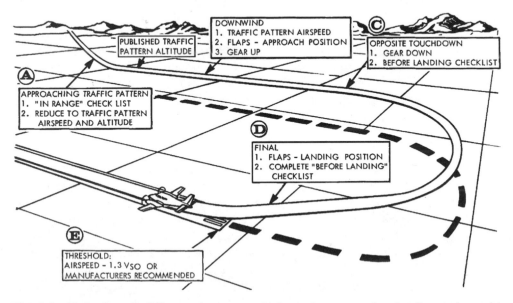

Fig. 9-9. *About the only difference between a single-engine approach to landing and a multi-engine approach to landing is a slightly higher approach, and waiting to lower flaps until the airport is assured.*

rect shutdown procedures. The sudden realization that this is not a simulated failure, often associated with an actual failed engine, often results in confused and hasty reactions.

When an engine fails during cruising flight, the pilot's first problem is to maintain a rational thought pattern, then to manage remaining altitude and airspeed to be able to continue flight to the point of intended landing. How far they can fly is dependent on the density altitude, gross weight, and obstructions. When the airplane is above its single-engine service ceiling, altitude will be lost, at least down to the single-engine service ceiling. Remember, the single-engine service ceiling is the maximum density altitude at which the single-engine best rate-of-climb speed will produce 50 FPM rate of climb. This ceiling is determined by the manufacturer on the basis of the airplane's maximum gross weight, flaps, and landing gear retracted, critical engine inoperative, and with the propeller feathered.

Although engine failure while enroute in normal cruise conditions may not be critical, it will darn sure get the pilot's attention. It is a good practice to add maximum permissible power to the operating engine before securing or shutting down the failed engine. If it is determined later that maximum permissible power on the operating engine is not needed to maintain altitude, then reduce the power and save the engine. I always tell my students, "One dead engine is enough already, let's baby the one that's gonna get us home."

However, with reduced power on the good engine, if the airspeed decreases too much, this could present a serious performance problem, especially if the airspeed should drop below V_{yse}. If this occurs, add whatever power is needed to the operating engine. In short, do not hesitate to use the power if you need it.

Altitude should be maintained if it is within the capability of the airplane. In an airplane not capable of maintaining altitude with an engine inoperative under existing circumstances, the airspeed should be maintained within ±5 knots of the engine-out best rate-of-climb speed (V_{yse}) so as to conserve altitude as long as possible to reach a suitable landing area.

After the landing gear and flaps are retracted and the failed engine is shut down and everything is under control, it is a good idea to communicate with the nearest ground facility to let them know the flight is being conducted with one engine inoperative. ATC facilities are able to give valuable assistance if needed, particularly when the flight is conducted under radar observation, which is nearly everywhere these days.

Good judgment would dictate, of course, that a landing be made at the nearest suitable airport as soon as practical rather than continuing the flight to a distant destination.

During cold-weather, engine-out practice using zero thrust power settings, the engine may cool to temperatures considerably below the normal operating range. This factor requires caution when advancing the power at the termination of single-engine practice. If the power is advanced rapidly, the engine may not respond and an actual engine failure may be encountered. This can be helped by closing the cowl flaps during periods of prolonged zero thrust on the idled engine.

This lack of heat on the idled engine is particularly important to remember when practicing engine-out approaches and landings. A good procedure is to slowly advance the throttle to approximately one-half power, then allow it to respond and stabilize before advancing to higher power settings. This procedure not only guarantees a steady supply of power, but results in much less wear on the engines of the aircraft.

Restarts after feathering require the same amount of care, primarily to avoid engine damage. Following the restart, the engine power should be maintained at the idle setting or slightly above until the engine is sufficiently warm and is receiving adequate lubrication.

ENGINE FAILURE CHECKLISTS

Although each aircraft must be operated in accordance with the POH, the following checklists are presented to familiarize the would-be multiengine pilot with the actions that are typically required when an engine fails.

Engine Failure During Cruise

1. Mixtures—As Required for Flight Altitude.
2. Propellers—Full Forward.
3. Throttles—Full Forward.
4. Landing Gear—Retracted.
5. Wing Flaps—Retracted.
6. Inoperative Engine—Determine.
7. Establish at Least 5-Degree Bank—Toward Operative Engine.
8. Inoperative Engine—Secure.
    a. Throttle—Close.
    b. Mixture—Idle Cut-Off.
    c. Propeller—Feather.
    d. Fuel Selector—Off.
    e. Auxiliary Fuel Pump—Off.
    f. Magneto Switches—Off.
    g. Alternator Switch—Off.
    h. Cowl Flap—Close.

9. Operative Engine—Adjust.
    a. Power—As Required.
    b. Mixture—As Required for Flight Altitude.
    c. Fuel Selector—As Required.
    d. Auxiliary Fuel Pump—On.
    e. Cowl Flap—As Required.

10. Trim Tabs—Adjust Bank Toward Operative Engine.
11. Electrical Load—Decrease to Minimum Required.
12. As Soon as Practical—Land.

Airstart (After Shutdown)

Airplanes without propeller unfeathering system:

1. Magneto Switches—On.
2. Fuel Selector—Main Tank (Feel for Detent).
3. Throttle—Forward Approximately One Inch.
4. Mixture—As Required for Flight Altitude.
5. Propeller—Forward of Detent.
6. Starter Button—Press.
7. Primer Switch—Activate.
8. Starter and Primer Switch—Release When Engine Fires.
9. Mixture—As Required.
10. Power—Increase after Cylinder Head Temperature Reaches 200 Degrees F.
11. Cowl Flap—As Required.
12. Alternator—On.

Airplanes with Propeller Unfeathering System

1. Magneto Switches—On.
2. Fuel Selector—Main Tank (Feel for Detent).
3. Throttle—Forward Approximately One Inch.
4. Mixture—As Required for Flight Altitude.
5. Propeller—Full Forward.
6. Propeller—Retard to Detent When Propeller Reaches 1000 RPM.
7. Mixture—As Required.
8. Power—Increase After Cylinder Head Temperature Reaches 200 Degrees F.
9. Cowl Flap—As Required.
10. Alternator—On

ENGINE-OUT APPROACH AND LANDING

Essentially, an engine-out approach and landing is the same as a normal approach and landing. Long, flat approaches with high-power output on the operating engine and/or excessive threshold speed that results in floating and unnecessary runway use should be avoided. Due to variations in the performance limitations of many light twins, no specific flightpath or procedure can be proposed that would be adequate in all engine-out approaches. In most light-twins, however, a single-engine approach can be accomplished with the flightpath and procedures almost identical to a normal approach and landing.

Fig. 9-10. *At this point, it's perfectly alright to exhale. You've earned it!*

During multiengine training, the pilot should perform approaches and landings with the power of one engine set to simulate the drag of a feathered propeller (zero thrust), or if feathering propellers are not installed, the throttle of the simulated failed engine set to idle. With the "dead" engine feathered or set to "zero thrust," normal drag is considerably reduced, resulting in a longer landing roll. Allowances should be made accordingly for the final approach and landing.

The final approach speed should not be less than V_{yse} until the landing is assured; thereafter, it should be at the speed commensurate with the flap position until beginning the roundout for landing. Under normal conditions the approach should be made with full flaps; however, neither full flaps nor the landing gear should be extended until the landing is assured. With full flaps the approach speed should be 1.3 V_{so} or as recommended by the manufacturer.

The pilot should be careful not to lower the flaps too soon on a single-engine approach. Once they have been extended it may not be possible to retract them in time to initiate a go-around. Most light-twins are not capable of making a single-engine go-around with full flaps. In fact, a single-engine go-around in light-twins under the best of circumstances is not something one wants to have happen. Do not place yourself in the position of having to go-around. Plan your approach and get it right the first time! (FIG. 9-10.)

10
Federal Aviation Regulations (Changes To)

THE FEDERAL AVIATION REGULATIONS (FARS) ARE TO THE PILOTS OF America what the Rules of the Road are to the drivers. Except if there were a tenth as many Rules of the Road as there are FARs, we wouldn't have nearly as many silly drivers to contend with on those roads. The FARs comprise many thousands of pages of government rules pertaining to just about every eventuality a pilot could conceive of when flying.

The FAA is responsible for the administration of the aviation rules and regulations which are divided into chapters the FAA likes to call Parts. The title of each chapter (Part) gives us a hint of what lies within. For instance, FAR Part 1 is called Definitions and Abbreviations and contains—that's right, definitions and abbreviations. FAR Part 61 is called Certification of Pilots and Instructors and contains the minimum requirements for the licensing of pilots, flight, and ground instructors. FAR Part 91 is called General Operating and Flight Rules and contains—you guessed it—general operating and flight rules which we pilots must abide by when we operate an aircraft.

Every so often, the FAA will make small revisions or update the FARs to conform with new information, safer policies, or merely to adhere to a new political climate. Usually this is done by rescinding a sentence or two and the insertion of new information.

However, FAR Part 61, Certification of Pilots and Instructors, the FAR that governs how pilots are licensed has just recently been given a complete overhaul by the FAA. Because of the importance of both FAR Part 61 and how it affects a soon-to-be Private Pilot or a pilot planning to upgrade to a Commercial Certificate or Instrument Rating, following is a review of the major changes of the new FAR Part 61.

FAR PART 61 REVISIONS

The long-awaited revision to regulations affecting pilots, flight, and ground instructors has been released. On April 4, 1997, the FAA announced the final rule, which became effective August 4, 1997. It significantly modified Part 61 of the Federal Aviation Regulations (FARs). The original proposal, which drew thousands of suggestions from pilots around the United States, included numerous changes including medical self-certifications for some pilots, separate instrument ratings for single and multiengine airplanes, a new balloon flight instructor certificate, and several new aircraft categories and classes.

As a result of the comments received by the FAA, many of the original proposed changes were dropped and others have been implemented in a highly modified form. In the paragraphs below I have outlined some of the more significant changes having widespread effect. Be aware that these are not verbatim FARs; I have paraphrased the actual regulations. In order to remain abreast of the complete changes to the FARs, and to view the format of their composition, it would be wise for you to refer to the actual FARs.

Applicability and Definitions (61.1)

This paragraph, which covers the hows and whys of FAR Part 61, has been greatly expanded. Among the new and revised definitions are aeronautical experience, authorized instructor, flight simulator, flight training device, pilot time, knowledge test, practical test, and training time. For instance, we old certified flight instructors are now called authorized flight instructors. I don't know if it pays any more or not, but that's what the FAA wants us to be called.

When studying the provisions required for certification contained within Part 61, it will be important to refer to this paragraph.

Definition of Cross-Country

The definition of cross-country is now contained entirely within paragraph 61.1 (applicability and definitions). For the purposes of meeting the requirements for private, instrument, or commercial certification, any point of landing must be a straight line distance greater than 50 nautical miles from the original point of departure. Another change is commercial and airline transport pilots may log cross-country time on any flight which includes a straight line distance of greater than 50 nautical miles from the original point of departure, regardless of whether a landing is made.

Logging of Flight Time (61.51)

Most of the changes to this section involve clarification of what have been long-standing points of confusion and disagreement. Here are the key items. Recreational, private, commercial, and airline transport pilots may log pilot-in-command (PIC) time whenever they are sole manipulators of the controls of an aircraft for which they are rated. Flight instructors may log PIC time whenever acting as an instructor.

A major change, student pilots may log PIC time when they are sole occupant of the aircraft, have a current solo endorsement, and are either undergoing training for a certificate or rating or building PIC time for a certificate or rating. Second-in-command time may be logged when more than one pilot is required either under the type certification of the aircraft or by the regulation under which the flight is being conducted. This includes when acting as safety pilot for another pilot under simulated instrument conditions.

Instrument Currency (61.57)

The new regulations eliminate the requirement for six hours of instrument time every six months to remain current. Instrument pilots must now accomplish the following within the preceding six months; six instrument approaches, holding procedures, and intercepting and tracking courses through the use of navigation systems. This may be accomplished in flight in either actual or simulated (with a safety pilot) instrument conditions or in an approved simulator or flight training device. The only alternative to these currency requirements would be for the instrument pilot to choose to accomplish an instrument competency check with an authorized instrument instructor in lieu of the approaches and holding.

Complex vs. High Performance (61.31)

The distinction between complex and high performance airplanes has been formalized. Separate training and separate instructor endorsements are required prior to acting as pilot-in-command. Complex land airplanes are defined as those having retractable landing gear, flaps, and a controllable pitch propeller. High performance airplanes are those with an engine of more than 200 horsepower. (FIG. 10-1.)

Recreational Pilot Privileges

In a change heavily lobbied for by aviation advocacy groups, the requirement to remain within the 50 mile distance limitation for recreational pilots has been lifted. Recreational pilots may now venture far and wide, providing they receive the same cross-country

Fig. 10-1. *This pressurized Cessna 210 meets both the high performance and complex specifications of FAR Part 61.31.*

training specified for private pilots. Other existing limitations remain, including no flying after sundown, or where two-way communication with ATC is required (tower controlled airports, etc.). Nor may the recreational pilot carry more than one passenger at a time in an aircraft with no more than four seats or engines greater than 180 horsepower.

Flight Instructor Renewal (61.197)

While the various methods for CFI renewal remains the same, CFIs will now be able to renew up to 90 days prior to their certificate expiration date with the new expiration date 24 months after the original date. This translation is a welcome relief to the CFIs who may have had trouble renewing on, or by, their expiration dates.

Retesting

For those of us who sometimes fall a little bit short when test time comes around, the FAA has deemed that retesting after the second failure of a written or practical (flight) test no longer requires a 30-day waiting period. This 30-day wait after failing a test for the second time was a good plan that didn't work well. The thinking was that by forcing the 30-day wait, the applicant would seek more dual instruction and therefore be better prepared for the failed exam. Sadly, about the only thing most of these folks did was wait the 30 days and then retake the test. It has been my observation, and I know there are exceptions, that any pilot who fails the same test twice usually won't do much better whether on the fourth or fortieth try.

HOURLY REQUIREMENTS FOR CERTIFICATION

Private Pilot (61.109)

Applicants for the Private Pilot Certificate still must have 40 hours of total flight time, of which 20 hours must be dual instruction and 10 hours must be solo flight training specific to the areas of operation listed in FAR 61.107. The remaining 10 hours may be either dual or solo. Up to 2.5 hours of the training may be completed in an FAA approved simulator or flight training device.

There must be a minimum of 3 hours of daytime dual cross-country instruction. Additionally, there must be 3 hours of night dual instruction which must include one cross-country of over 100 nautical miles total distance and at least 10 takeoffs and landings to a full stop. The option of not conducting night training and receiving a "night flying prohibited" limitation no longer exists. The only exception is for pilots training and residing in Alaska.

The Private Pilot applicant needs a minimum of 3 hours of dual instrument flight training. This represents a significant change; previously there was no specified minimum instrument training. The total instrument time was left to the discretion of the flight instructor who had to declare the student was proficient at the instrument flight maneuvers required for the private license.

The minimum of 3 hours of dual flight instruction in preparation for the practical test, within 60 days prior to that test, remains the same. This regulation is designed so the applicant for the license will have had some recent formal dual instruction prior to the checkride.

The solo cross-country flight time has been reduced to 5 hours total flight time. One solo cross-country must be at least 150 nautical miles total distance, must include one non-stop leg of at least 50 nautical miles, and must include full-stop landings at a minimum of three points.

The Private Pilot applicant must also have made at least 3 full-stop landings at an airport with an operating control tower. These full-stop landings may be dual or solo and are designed so that all private pilots will have had at least a minimal exposure to a tower-controlled airport prior to their receiving their license.

Instrument Rating (61.65)

The big change in the instrument rating is that the previous requirement for 125 hours of total flight time has been eliminated. This change will allow newly certified private pilots to begin training immediately for an instrument rating. Applicants are still required to have 50 hours of pilot-in-command cross-country (again, see definition above). However, PIC cross-country time obtained as a student pilot will now count along with all PIC cross-country flight time logged after receiving the private license.

Fig. 10-2. *A favorite instrument trainer of students and flight schools alike is the Cessna 172RG.*

Forty hours of actual or simulated instrument time must logged to include a minimum of 15 hours of dual instruction, with at least 3 hours of that 15 hours coming within 60 days prior to taking the practical test (See FIG. 10-2). Up to 20 hours of instrument instruction in an approved simulator or flight training device may be counted.

Instrument cross-country dual instruction must include at least one flight of at least 250 nautical miles along the airways or direct as cleared by ATC. Additionally, there must be one instrument approach at each airport and these approaches must use three different kinds of navigation systems.

Commercial Pilot (61.129)

Total flight time required for commercial pilot applicants remains at 250 hours (of which 50 hours may be in a simulator or flight training device). This must include 100 hours in powered aircraft (including 50 in airplanes), 100 hours PIC time (including 50 in airplanes), and 50 hours of PIC cross-country (see definition above). However, some very significant additional requirements have been adopted.

There must be twenty (20) hours of dual instruction on the areas of operation specific to commercial certification (listed in 61.127) which must include the following:

- 10 hours of instrument instruction.
- 10 hours in a complex or turbine-powered aircraft.
- One day-VFR cross-country flight of at least two (2) hours and more than 100 nautical miles from the original departure point.
- One night-VFR cross-country flight of at least two (2) hours and more than 100 nautical miles from the original departure point.

There must also be three (3) hours in preparation for the practical test within 60 days preceding and ten (10) hours of solo flight (this means sole occupant of the aircraft) which must include at least one cross-country flight of at least 300 nautical miles, with landings at a minimum of 3 points, one of which is at least 250 nautical miles from the original departure point.

There must also be five (5) hours of night-VFR which must include at least 10 take-offs and landings at an airport with an operating control tower.

This sums up the major changes to FAR Part 61. If you are on the path to certification, it would be wise for you to pick up a copy of the current version of the FAR/AIM at your local airport or bookstore. As its title implies, the FAR/AIM contains the major FARs and the total Aeronautical Information Manual. This compact $10 or $12 book has more information in it than one could ever absorb, comes out yearly, and is literally the Bible of aviators.

11
Aviation Careers

STUDIES SHOW THAT THE MAJORITY OF PEOPLE DO NOT LIKE THEIR chosen profession. Further, most of us dream of being something, or someone, else. Farmers want to be teachers and teachers want to be accountants. Nearly every person I know can't wait for Friday night and age 65. Not so with most professional pilots. Chances are, a professional pilot wouldn't change places with anyone. The phrase most often heard when a pilot describes his job is, "I have the best damn job in the world."

Flying holds a certain fascination for almost everyone. In over three decades as a pilot, I have spent more time at school, parties, and family reunions talking about flying than all other subjects combined. And I am rarely the originator of the conversation. It seems that when people find out I fly for a living, that's all they want to talk about. I suppose I have been told by at least 10,000 people, "I've always wanted to learn to fly since I was little. I guess someday I'll get around to it."

Flying intrigues many people. In fact, it probably weaves its way through the daydreams of most people, but not to the extent that they are willing to drop everything and pursue it. In short, that's not what they really want to do.

And then there are the true fliers. For we pilots, there is absolutely nothing we could be doing for a living that would come close to flight. To us, the worst flying job on the planet would be 10 times better than whatever comes in second place.

If you feel you would like a career in aviation, you quite likely feel it deeply. Few people are drawn to flight by a superficial urge. It seems to be a longing—an almost insatiable desire to fly—which is not diminished by neglect. If I have not flown for a period of time, I feel, as do most other pilots, a sort of inner anguish. And it can only be satisfied by flight.

If you love flying, you'll know it and no one will have to convince you to go get that flying job. In fact, no one will be able to stop you. To a pilot, flying is the ultimate career, and the flying jobs are there! All you have to have is a real desire, a modicum of ability, and the tenacity to stick it out during the early years. For the competent, dedicated pilot, the sky really is the limit!

AIRLINES

An airline career seems to hold the most fascination to the most people. Without a doubt, this career has the most visibility, good and bad, and has the highest earning potential of all the current legal aviation careers. Additionally, there seems to be little doubt that the airline industry will continue to grow at a steady pace well into the twenty-first century. This, coupled with the natural retirement rate, should guarantee a need for quite a few airline pilots over the coming decade and beyond. Some people have gone so far as to predict the need for 3,000 new airline pilots per year well into the next century.

I believe these figures might be a bit optimistic since our present airport and Air Traffic Control system is near its capacity. Until we build more airports, or expand our present ones, I cannot see a dramatic rise in the need for airline pilots. And I haven't a clue as to how ATC can handle many more aircraft than it now handles and maintain the same margin of safety in the traffic flow.

Another factor that will figure into the potential airline job market is the advent of two-man crews. Every airliner built in the United States today is crewed by two pilots instead of the traditional two pilots and a flight engineer used for so many years in the old 707s, DC-8s, etc. (See FIG. 11-1). This is the price we pay for our technical advances. The sophistication of EFIS (electronic flight information systems) instrument panels has so centralized pilot information that it has eliminated the need for one human. This also cuts by one-third the need for manpower.

Up until the early 80s, if a pilot was fortunate enough to be hired by a major airline and gathered a few years of seniority, he had it made. Then, seemingly in a few short months, it all changed. It changed with the Federal deregulation of the airline industry. For the first time in their history, the airlines were free to do as they pleased.

The federal guidelines were gone, and what followed was a frenzy beyond the wildest dreams of most aviators. There was a wellspring of new carriers, mergers, buy-outs, takeovers, fare-wars, restructuring, and ultimately, bankruptcies. The air carriers became convinced they would have to grow or be consumed. They were right.

A new idea sprang onto the scene—commuter airlines. Nearly every city with an airport had a commuter airline. Never mind that it had no passengers and made no money, Smallsville, U.S.A., had an airline. One by one, these commuters were bought, merged with, or simply taken over.

Fig. 11-1. *This Boeing 767 is representative of the type flown by the majority of U.S. air carriers.*

This growth brought on a need for pilots, which triggered a surge of hiring by the airlines, especially the commuters. The problem was that there was a lack of qualified pilots from which to choose. The commuters' solution to this problem was to hire pilots that not only weren't dry behind the ears—they'd never been wet. This caused a great cry of concern from seasoned pilots and much of the flying public. The concern was that safety was being sacrificed for the sake of routes. And several very serious accidents were directly attributed to lack of flight crew experience.

All of these buyouts and mergers left the airline pilots in a state of flux. A senior captain with one airline today could very easily be wearing a different hat with a low seniority number the very next day. Often, pilots with many years of seniority were furloughed indefinitely when their line was merged with another. It all depended on which line you happened to fly for. Some pilots lost everything when their financial future that had looked so solid turned sour. In short, it would have been easier to nail Jell-O to a tree than to predict the future of an airline pilot in the 80s. It was wild.

About 1990, as quickly as it had begun, the hiring boom was quieted. The major airlines began a more methodical, safety-paced approach. Hiring minimums began to take on the look of yesteryear. More experience would be needed to qualify for that coveted seat aboard an airliner once again. The quick trip to the majors was over, at least for now.

Then, about 1995, the airlines began to notice something they had not prepared for. The economy had been spiraling upward for several years, the average person had more disposable income, and people were flying to their vacation spots instead of driving. The airlines, leaner and smarter than before, were beginning to make money. And looking into the twenty-first century, there appears to be no end in sight.

The point I'm attempting to convey with this brief history lesson is that the airline job is not all the sweetness and security that so many would have you believe. It is a tough, often disappointing, career that is not for the timid. If you desire to be an airline pilot, you had better go in with your eyes open, your head on straight, and your proficiency unquestioned. If not, it will eat you alive.

This is not to say that there aren't any great jobs flying for the airlines. For those who do make it into the left seat, and manage to remain high in seniority, it can be a very rewarding career. How many people do you know that work three days and then get four days off? And the pay's not bad, either.

I asked many airline pilots what the most important prerequisite was for landing a job with a major airline. Their answer: experience, experience, experience. Naturally, the airlines would prefer an ex-military pilot who has already been trained and could come to them with several thousand hours of heavy multiengine jet time. If you want a job with the airlines, this is as close as I can come to guaranteeing you a seat. I know of no airline that turns away pilots with an adequate amount of military multiengine jet time.

Interestingly, when I asked several airlines about the greatest asset they sought in a pilot candidate, the resounding answer was attitude. The close work environment of modern airliners demands a cohesiveness and interpersonal ability to work and play well with others that the airlines find absolutely invaluable in today's pilot applicant. Several airline human resource managers told me that attitude had in fact replaced experience as their number one hiring priority.

Some airlines are not enamored by the fighter pilots. It seems that some fighter pilots have a single-minded, cover-your-own-butt mentality. And who could blame them? It is reasoned that this could interfere with the teamwork necessary in an airline cockpit. While this is certainly not a given, it is a consideration, and most airlines seem to prefer the four-engine types.

For those of you not in the mood to give those years to Uncle Sam in exchange for receiving invaluable training, the next best path to the airline cockpit is to somehow build flight time. The airlines prefer that you have a couple thousand hours of flight time, preferably turbine. Their dream pilot would have a degree in aeronautics, 3,000 hours of flight time, of which about 2,000 hours was in turbine-powered aircraft, and be about twenty-eight years old. That's their dream, but they will settle for a bit less.

The personnel I talked with told me that a college degree is as important as it ever was. In fact, one airline HR person told me flatly that a full 99.9 percent of their new-hires had a four-year degree. So naturally, if you have a degree, you will be considered (read that *hired*) before someone with the same qualifications who does not have a college diploma. In reality, a degree is very helpful both at the time of hiring and later during promotion time.

There are many ways to build the flight time desired by the airlines, and no one system works equally well for everyone. (I was told one point, pretty plainly. The airlines do not care if you have 8,000 hours in your Cessna 172; they want turbine time.)

An acquaintance of mine took the following route to his airline seat. He soloed at the age of 16, obtained his commercial when he was 18, and then went to college to earn his bachelor's degree. During his college summers, he ferried new aircraft to dealers for a large aircraft manufacturer and managed to build up another 1,000 flight hours. More importantly, he got paid to fly. Smart!

Upon graduating from college, he managed to land a corporate job flying a King-Air for a power company. He spent eight years working for this corporation, during which time they conveniently added a Sabreliner and a Cessna Citation. During these years, he flew an additional 3,000 flight hours, all turbine.

He was now ready, at the ripe old age of 30 to apply to the airlines. He found a place with a commuter airline and flew with them for two years before making the jump to the majors. He flew five years as a First Officer (copilot) on a DC-9 and currently flies right seat on an A-320 Airbus. He is 37 and should make captain in the next five years.

That's all well and good, you say, but does it pay? You'd better believe it. As I said, if you can land the job and manage to remain with a solvent company for six or seven years, you will be well compensated. Our friend in the above paragraph will make $98,110 per year as of August 31, 1998. And he is a third-year copilot. A 10-year captain flying a 747 will make $202,500 for the same year. In fact, this year the lowest paying flying job with the airline I am quoting will be $59,880 for a second officer on a 727. You could live on this to start, couldn't you?

Also, the major airlines offer great benefit packages including life, accident, and health insurance, profit sharing, stock options, annuities, and so on.

The pros and cons depend on to whom you happen to be talking. The obvious negatives are the takeovers, which cause loss of seniority, and the possible layoffs. In this volatile time in the evolution of the airlines, this seems to be the main complaint of all airline pilots. Following this on the negative side are union disputes, long hours (like 16-hour days), and being away from home. The latter was also mentioned as a plus. I guess it depends on your priorities.

On the plus side, money, money, and money were mentioned in first, second, and third place. They were followed by free travel, short work weeks, and excellent retirement plans as solid reasons for working with the airlines.

Speaking of retirement, airline pilots *must* retire from the flight controls at the age of 60. They can remain as flight engineers but cannot man the two front seats any longer. This is a mandatory retirement age as decreed by the feds, and they seem unwilling to discuss it.

The reason for this arbitrary age of 60 seems to be predicated on the FAA's assumption that a pilot of 60 is more likely to die suddenly than a pilot of 25. This is known in aviation circles as *FAA intelligence*. There is a fairly large push among pilots to set the retirement age back to 65, saying that the experience gained by this age is irreplaceable. However, most airlines seem to favor the 60 rule, probably for financial reasons. The pilots who remain in place to age 60 are usually making some serious money. Anyway, an age for mandatory retirement had to be chosen, and for now, it remains 60.

AIR TRAFFIC CONTROL

If you don't want to fly for a living but desire to remain close to aviation, one of the most attractive nonflying vocations could be a career with ATC. ATC, which includes control towers, air route traffic control centers, and flight service stations, is run by the government—the FAA to be precise. The folks in ATC work with all types of aircraft on a daily basis, assisting their every move from clearance delivery before engine start to closing flight plans after shutdown. Given the ever-increasing number of aircraft sharing our seemingly shrinking airspace, job security should be solid for the foreseeable future. (See FIG. 11-2.)

In order to land a position as a controller, you must first be placed on the Federal Register. The register is a list of all prospective employees the government utilizes to fill vacancies as they occur. You are placed on this list by successfully passing a government test designed specifically for ATC hopefuls, called the *ATC Operators Test*. This written test is scheduled for various parts of the country at periodic intervals, or for specific areas when the need for ATC personnel is apparent. You can obtain a current test schedule by contacting your nearest Federal Office of Personnel Management or by calling your nearest FAA Flight Standards District Office.

Once you pass the written test, the FAA will screen you for physical, psychological, and security reasons. If you clear this, you are summoned to Oklahoma City for four months of initial training at the FAA's training facility. The high scores from the Oklahoma City schooling are given the choice of entering either the tower or the enroute (center) phase of the ATC system. Either way, you will serve two years of apprenticeship before being certified as having reached ATC FPL (full performance level) status.

ATC personnel work a standard eight-hour day, 40-hour week. And since the stress of separating numerous high-speed aircraft can be counterproductive, their contract calls for a short break once every two hours.

Fig. 11-2. *An ATC controller is often as involved in the flight as the pilot.*

ATC salaries are figured by several criteria, including the traffic count at the facility worked in comparison to all the others. For instance, ATC control towers are divided into five levels based on traffic count. An entry-level controller based at a level-one facility (light traffic) will make about $24,000 in 1998. An FPL controller working a level-five tower (very busy) will earn about $58,000.

When you add in government health and life insurance benefits to this, you begin to see possibilities of a career that provides a pretty comfortable living.

Additionally, all ATC personnel earn paid leave at the rate of a half day off per pay period at the entry level. This rate grows proportionally until, after 15 years of service, they earn one day of leave per pay period. This figures out to between 13 and 26 paid personal days per year in addition to regular days off and 10 paid federal holidays per year. Controllers also accumulate a half day of sick leave for every pay period, or 13 days per year. This all adds up to a very tidy seven weeks off with pay for a veteran controller. Not bad.

Now, an age-old question: Should an air traffic controller also be a pilot? Should they not know something about what it is pilots do since, after all, pilots are placing their lives in their hands. Some say no; other say absolutely; and some are ambivalent towards the whole subject. One camp says that an ATC person who can relate to the pilot and identify with the problems faced up there in three-dimensional airspace is better prepared to help the pilot overcome obstacles to safe navigation.

Another camp says that this identity with the pilot and his or her plight is the exact reason a controller should *not* be a pilot. They say the controller cannot, and therefore should not, attempt to fly the plane. They maintain that the controller should do the controlling and the pilot should fly the plane.

I believe the best solution, as usual, lies somewhere in between. Given the choice, I will take a controller who flies, and yet knows both his and my limitations. This controller will know enough about the problems I am facing in the aircraft to be of great help to me, thus allowing me to do my work in the most efficient manner. Further, this type of controller can most likely be counted on to help, and perhaps equally important, will not compound an already difficult situation by issuing an impossible instruction.

The comments, pro and con, I heard most concerning a career with ATC had to do with opposite ends of the same thought. The consensus was that the worst part of a controller's job was either 1) stress, or 2) boredom. Imagine that! They seem to be either too busy, or not busy enough. I guess pilots had better become more skilled at spacing arrivals so as to spread them more evenly over the day. Come to think of it, that is not such a bad idea.

The consensus was that the best part of the controller's job description was the chance to interact with and aid the pilots. I saw the warm glow of satisfaction in the eyes of more than one controller as they described how they had helped a pilot, who had flown past his fatigue or experience level, culminate his flight successfully. And the pride was all the more evident if they hinted that they had to bend the FAA's stringent rules in order to discreetly suggest a better way to a tired or frightened pilot. That is a kind of job satisfaction few of us will ever have the chance to enjoy.

ATC is a viable career with much growth potential and job satisfaction. Most controllers I talk to are absolutely in love with their profession. If you would like to be around aviation, and yet not fly for a living, this could be the career you have been looking for.

FLIGHT INSTRUCTION

"What's Al do for a living?" "Oh, he's 'just' a flight instructor. As soon as he gets enough hours logged, he's gonna move up to corporate flying."

I wonder how many times I have heard this pathetic exchange between two otherwise intelligent people. Some believe that flight instruction is the poor step-child of aviation. They think of flight instruction as the sort of flying job that should make you hang your head when someone asks you what you do for a living; the sort of job you take to build flying time so you can go get a real job. Sadly, these naive thoughts seem to have become "tradition" in some areas of aviation. But they are only the traditional thoughts of the ill-taught and the unknowing. The rest of us realize the absolute undeniable importance of the flight instructor and know that they are vital to all who desire to learn to fly (See FIG. 11-3). The official position of the governing body of aviation, the FAA, is that the CFI is the backbone of aviation.

It's true that over the years many pilots have used flight instruction as a method of building flight time. Far too many CFIs have used the position to simply ride along with their students, building time instead of teaching. Those of us who have chosen to make instructing our career hold the aforementioned in the utmost contempt. And some of us who are in a position to hire these joy-riders as flight instructors, don't.

Fig. 11-3. *One of the most important persons in your flight career is your primary instructor.*

I'll tell you a secret. You will fly more like the instructor who teaches you early flight (Private Pilot) than like *anyone else* you will ever fly with. It's only natural. You talk like the people who taught you to talk and walk like the people who taught you to walk, don't you? And you drive like the people who taught you to drive, don't you? You bet you do. You learned the basics from watching and copying the first role model you ever had. And so it is with flight. I hope you had, or will have a great primary flight instructor because you will be a clone of his or her ways, good or bad. And if you are contemplating a career as a CFI, remember this, and teach accordingly.

In flight instruction, as with all vocations, not all who teach are suited to the job. A flight instructor's duties include the ability to teach, counsel, coach, coddle, trick, exhort, and support the student in one of the most difficult teaching environments imaginable. The cockpit is a tough classroom, and it takes a special personality to maintain self-control as a student repeatedly tries to kill you. This gets even more problematic as you and the student begin to develop the natural bond that comes with many hours shared. It is difficult to critique accurately and subjectively anyone with whom you have developed a fondness.

For this reason, I try to never get too close to the people I train to fly—for their sake. I don't want to be in the position of endangering their lives because I like them so much that I might not want to risk hurting their feelings. I hope I wouldn't do this, but I have seen it happen to others. When my son learned to fly, I sent him to the best instructor I knew rather than risk the tendency to overlook problem areas. My son says he wouldn't have let me train him anyhow. He says I would have been an unmerciful tyrant who would have settled for nothing less than perfection. He's probably right. Either way, I believe it is best not to train people you are close to.

I'll tell you another secret. If you are a pilot of better than average abilities, no matter where you work, you will be asked to teach. It is a high honor, indeed, to be asked to teach. Flight schools, colleges, universities, the armed forces, airlines, the FAA, and many corporate flight departments have their own flight instructors. And they usually don't offer the position of training the pilots they will have to depend on to the underqualified or the inept. I offer as an example the U.S. Navy "Top Gun" Aerial Warfare School. These guys are "just" flight instructors, only they happen to work for the Navy.

It is a solemn responsibility to accept the flight instructor task. I consider it the highest form of compliment when someone is willing to entrust me with the flight training of their son, daughter, husband, or wife. It is my honor to teach them a craft that few attempt, and at which even fewer excel. It is hard, sometimes frustrating work, but it's always rewarding.

If you would like to become a flight instructor, you will first need to attain your Commercial Pilot Certificate with Instrument Rating. You must then pass two written exams: Fundamentals of Instruction and Flight Instructor Airplane. With the writtens safely in the bag, all that remains is to complete the flight time and pass the oral and flight check. While all this only fills a couple of lines here, it will generally take about two years to complete these requirements as a full-time student starting from the beginning. It can also be done several other ways, such as in your spare time (which could take forever) or at one of those accelerated flight schools. My personal preference would be to opt for either of the first two choices. I am not a fan of forced training.

Chapter Eleven

Now, you're ready to work. But where? The ranges of opportunities for employment as a new flight instructor are quite varied. Some new CFIs will be asked to work for the flight school that trained them. This invitation is usually predicated upon being in the above average range I talked about previously. Others will have to use their initiative and read the trade magazines to learn where work is available.

If you desire to work in a certain area of the country, be advised that the areas that are described as attractive will often pay less than their counterparts that are perceived as less desirable. What I'm saying is that you will most likely receive a smaller salary in the warmer, tourist-laden climates like Florida or California than you will in upper peninsula of Michigan. The main thing to remember is that there are opportunities in nearly every area. All you have to do is decide where you want to live, and then go for it. Somewhere nearby there is always work for a good CFI.

In order to attain serious earning potential as a CFI, you will have to land a position with one of the large fixed-base flight schools, a university, corporation, or the military. These training institutions usually offer a steady salary, great benefits, and a constant stream of students.

For instance, a CFI I am very familiar with began his career in a small fixed-base operation in Texas. He worked there for a year and a half, flying over 1,000 hours as an Air Force ROTC instructor. During this time, he also gained a lot of experience crop dusting, another high-tech, high proficiency area that would prove invaluable to him in the near future. (See FIG. 11-4.)

One day, an acquaintance told him of a large university flight school that was interested in initiating an aerial application curriculum. He applied and was accepted. This combination of luck, talent, and proficiency culminated in his teaching for 20 years at the university. For him, it was the best flight instructor job he could have asked for. He received a very livable salary, had great medical and life insurance, annuities, etc., and worked only nine months per year. During this 20-year period, he flew approximately 360 to 370 hours per year, or about two hours per day. Oh, yes, he also taught ground classes three afternoons a week for two hours.

The university connections allowed him to work with many other areas of aviation such as the EPA, the FAA, and various state and local officials, and to speak at large gatherings at other universities about his favorite subject: aviation safety. The FAA appointed him an Accident Safety Counselor and a Designated Pilot Examiner, one of only 1,700 worldwide. Another attractive bonus of college employment was the opportunity to attend classes without cost. This allowed his two children to obtain a nearly cost-free college education.

When the 20-year mark rolled around, this CFI retired with a hefty teacher's retirement account and a handsome annuity paid for by the university. He then moved to another college, where he went into administration and presently earns about twice as much money as he did at the previous university. This man is just over 50, has flown many thousands of hours, seen sights most men only dream of, was home nearly every night, loved every minute of his career, and is very close to being financially set for life. Who would have thought this possible for a person who began as, and still remains, "only" a flight instructor? Believe me, it can be a wonderful career.

Fig. 11-4. *Flight instruction is essential when checking out in a different type of aircraft.*

CORPORATE PILOT

"I just got back from 18 days in Europe," a corporate pilot friend said to me one day. "We were scheduled to be over there for six days, but business was better than they expected."

Now, this friend flies a Falcon 900 tri-jet, an extremely fine aircraft, and I imagine if I had to go to Europe, this would be as good a way to go as any. But 18 days for a six-day trip? Not me. He, on the other hand, wouldn't trade places with any other pilot in the world. He loves it!

Depending on your point of view, corporate flying has to be either the greatest thrill of all times, or the worst flying job imaginable. If the idea of rarely knowing where you will be going on a trip, or when you will return, sets you on fire, this is surely the flying job for you. The corporate world of flight carries with it some of the most interesting flying imaginable sometimes coupled with the most irregular hours that one could imagine.

On the other hand, many corporate flight departments are handled on a strict schedule. Crews know where and when they will fly, and for how long. They are run as if they were an airline. Many flight departments employ dispatchers who coordinate the scheduling of company aircraft with airlines, car rentals, etc., so the boss can minimize his travel time. With fewer and fewer "mom and pop" corporate aircraft flying, and profitability factoring heavily into the decision of whether to keep the aircraft or not, utilization has become a very important piece of the corporate picture.

I have many friends who are employed as corporate pilots, and they all, to a man, love their work. They like the fact that unlike airline pilots, they never get bored with a route, because they never know where they will be going next. They also seem to gain some odd sense of pleasure from this lack of regimen. In short, with most corporate flying, you fly to where the business is located. Sometimes this will take you to a certain city repeatedly, and sometimes it will be months between seeing a familiar area. You literally go when and where the boss wants to go and remain until he is ready to go home. You are, after all, carrying the corporation's finest in the back.

This diversity of time and locale seems to heighten the enjoyment of the true corporate pilot. The controlled chaos of juggling a schedule that matches business needs with aircraft schedules, weather delays, crew duty times, and scheduled and unscheduled maintenance is not a job for the fainthearted or the easily distracted. It takes a person with a high degree of flexibility to manage a career as a corporate pilot. (See FIG. 11-5.)

If a career in corporate aviation interests you, begin with your education. A degree in aviation with a minor in business will be the best two friends you could have at interview time. Your next best friends will be a couple thousand hours in your logbook, with about 500 hours multiengine time, and an A&P. Since many corporations like a combination pilot/A&P, an A&P certificate wouldn't hurt, if you can muster the time, energy, and expense in addition to your other schooling.

Whoa, wait a minute! This is getting to be real serious, isn't it? You bet it is. Don't think for one second that a Fortune 500 company is going to throw you into the front office of their Gulfstream IV just because you think it is something you'd like to do. With corporate aircraft in the 20-million-dollar range, and given the very nature of the cargo (executives) in the rear, it takes a very skilled person to break into the corporate ranks. This is some serious flying, and you will have to prepare accordingly. In fact, most corporate flight departments are run to airline standards. They legally operate under FAR Part 91, but boost it to the FAR Part 121 (airline) level for their flight operations.

Don't let these requirements slow you down. Just chart an intelligent course, build some flight time, preferably multiengine, and be prepared to settle for a copilot position for a few years. I know of several pilots who fly King-Airs for corporations who are not even typed in the aircraft. They signed on as copilots, and they wash the aircraft and even sweep up the hangar. They don't care. They know that one day the King-Air will turn into a Falcon 900 and they will be in the left seat. It's called *paying your dues*. After all, you can't be handed the knowledge required to fly a corporate jet to far-off places and arrive safely. You must start at the beginning and learn the trade. That is the primary reason for the corporate pilot's paycheck—the boss has faith he will get to and from the desired destination in the greatest comfort and safety. For that, he wants a pro. (See FIG. 11-6.)

The rate of pay varies from company to company, but a beginning pilot with a good background can start with a fair-sized corporation in the $26,000 to $35,000 range. Most corporate flight jobs peak at approximately $80,000 per year. The standard life, accident, and health insurance is usually a given, and many companies offer a bonus incentive program driven through corporate earnings. When you figure that many corporate pilots fly only three or four days per week, this begins to become pretty attractive. Most corporate pilots have about 120 hours of duty time per month, out of which they will fly about 40 hours.

Before you get too excited, let's look at what some pilots said about the good and bad of corporate flying. The most-often mentioned downside seemed to be the long

Fig. 11-5. *The corporate cockpit can be a busy place to work and many corporations now hire only proven team players.*

Fig. 11-6. *Not 9 to 5, maybe, but who cares?*

days. Some pilots told me of days that begin at 4:30 A.M. and last until 7:00 P.M., or longer. The flight time on a day such as this is usually in the five-to-six hour range with the rest of the time spent on preflight and postflight duties or just waiting. Though the corporate pilot usually flies just three or four days per week, the average duty day lasts about 12 hours.

Independence, or the feeling of being self-employed without the financial responsibility, was one of the most used responses for the positive side of the corporate career. Variety in aircraft flown as well as destination was a close second. Good pay, the feeling of being needed, and equal administrative treatment rounded out the pluses.

The one point on which nearly all corporate pilots are united is that their career area will grow steadily more competitive as economics force the elimination of many flight departments in the near future. Quite a few corporate crews will find that they have lost their aircraft, as companies trim to their respective fighting weights for the predicted monetary competition in the next century. This will most likely require the successful applicant to be more skilled as the decade unfolds. There will be jobs, but there will also be competition for those jobs.

If your desire ranges in the corporate area, go for it. It has been my experience that the proficient, prepared pilot is always employed, and sooner or later, that employment is in the desired area.

MILITARY

To many aviators, and would-be aviators, a career in military aviation is the very pinnacle of success. Just stop and think for a moment: How many pilots ever get the opportunity to fly an F-15, F-18, or C-5 Galaxy? If you make the grade as a military aviator, these are but a few of the aircraft you could potentially be assigned to fly. The opportunities are as boundless as the sky. You could catapult from the deck of a carrier in an F-14, pilot a C-5 to drop a load of cargo into some far-off country about which you have only read, or fly a low-level bomb run in an FB-111. (See FIG. 11-7.)

Since the map to the pilot's seat for all American military operations are nearly identical, I have arbitrarily chosen the Air Force to use as an example here. While the basic steps will usually remain unchanged, sometimes the requirements can change quickly. In order to get up-to-the-minute information for the military branch that interests you the most, I suggest you see your local recruiter. For instance, the Navy has only recently relaxed their vision requirements from the rigid 20-20 or better, which has been required for years.

In order to be considered for Air Force pilot training, you must be a college graduate, not less than 20 1/2 years of age nor more than 26 1/2 when you enter pilot training. These are the absolute musts.

The first step in earning your Air Force wings is to be commissioned as an Air Force officer. There are three paths to arrive at this goal: 1) You can enlist in the Air Force ROTC (Reserve Officer Training Corps) while attending one of the 600 participating colleges and universities across the United States. You simply enroll in the aerospace studies course at the time you register for your other freshman courses. Then, during your college years, the Air Force ROTC will prepare you for duty in the Air Force after graduation; 2) You can be chosen to attend the Air Force Academy. The Academy is a fully accredited four-year college, and upon graduation you are automatically qualified for your commission; or 3) You can be commissioned by successfully graduating from the Officers Training School (OTS), a 12-week course offered by the Air Force at Lackland AFB, Texas. In OTS, people who already have earned their college degree are training to become Air Force officers. (See FIG. 11-8.)

Prior to entering the undergraduate pilot program, all prospective Air Force pilots have to have had flight training in ROTC or at the Academy, or possess a civilian private pilot license. Those entering training through OTS will go through a Flight Screening Program (FSP), during which time they must advance to solo in a T-41, a military version of the Cessna 172.

Upon graduation from ROTC, OTS, or the Academy, you will enter Air Force undergraduate pilot training. This is where the fun really begins. During these 49 weeks, you will receive 176 hours of intensive flight training divided between the T-37 and the T-38 jet trainers. The initial 4 1/2 months consists of training in the T-37, a subsonic jet

Fig. 11-7. *F-15 Eagle.*

Fig. 11-8. *Afterburners on, an F-15 streaks skyward.*

trainer manufactured by Cessna. In this phase, you will be taught aerobatics, instrument and formation flying, and navigation.

After completing the T-37 phase, you train for an additional six months and 101 flight hours in the supersonic, twin-engine, T-38 "Talon." In this phase of training, you will delve more deeply into aerobatics, navigation, and instrument and formation flight while flying this advanced trainer, sometimes at altitudes as high as 55,000 feet.

After completing this final phase of undergraduate pilot training, you will receive your Air Force silver wings. You are now ready for advance training in the operational aircraft of the Air Force flight line. You will be selected to fly a type of aircraft based, for the most part, on the present needs of the Air Force. However, some latitude is given in allowing graduates to choose their desired aircraft. The training period for the operational aircraft is usually an additional three to six months.

The rate of pay in the Air Force, while not up to par with most civilian industry, is pretty good. It's good enough that a fair amount of Air Force pilots choose to make a career out of military aviation. The Air Force brochure says that it is "a unique way of life with an excellent compensation system to support it."

During training, a single, second lieutenant receives a base pay of $1,862 each month and pays income taxes on this amount. Upon graduation, as a captain, the base pay jumps to $3,062 per month. Additional bonuses are free housing in Junior Officer's Quarters and a subsistence (food) allowance. A major with 12 years service earns a base pay of $3,149 per month, $706 for housing, and $127 for subsistence.

And, oh yes, flight pay. Flight pay ranges from $125 per month for the first two years through a scale that peaks during years seven through 17 with a monthly flight pay of $650. This is in addition to all other salaries.

Additionally, you receive nearly free medical and dental care for you and your family, 30 days of vacation annually, discount shopping at the base exchange and commissary, and base housing with free utilities.

Well, there you are. A college education, OTS, and about a year and a half of additional training buy you a career that begins with an eight-year commitment to the Air Force. That's really not bad considering the government invests over a million dollars in your training—and you get paid while you train.

There is one more consideration I want you to remember before you run awestruck for your nearest recruiter: In real life they shoot real bullets, so do you—and the bullets go both ways. That's the down side of a military career. If you can handle that, go get it; it can be a great way of life!

OTHER POSSIBILITIES

The career areas I have discussed are merely the tips of a vast aviation complex we have to choose from. The uses for aircraft and pilots seem to grow with each passing day. The majority of the areas I talked about are for fixed-wing pilots, and I didn't touch on the career possibilities for pilots of helicopters, balloons, and blimps. I also didn't mention such nonflying areas such as A&P mechanic, flight dispatcher, steward, airport personnel, airport management, and support areas such as FBOs (fixed base operators), fuel

Fig. 11-9. *Without A&P mechanics to service the aircraft, we wouldn't need pilots to fly them.*

retailers, etc. These areas make up a vast portion of aviation, and we couldn't get along without any of them, but I really wanted to deal with the area I know best—the cockpit and the opportunities that lie in the front of the fixed-wing aircraft. (See FIG. 11-9.)

There are many other flight positions that you could consider in your aviation career search. A few that come quickly to mind are fish spotting, fire bombing, crop dusting, sightseeing tours, aerial photography, pipeline patrol, powerline patrol, and air taxi and charter. I'm certain I left out a few, but I believe you can see from this the great variety of potential aviation careers. (See FIG. 11-10.)

A career in aviation is also, no matter what or where you fly, the absolute best way to make a living that exists on the face of this earth. At least to us pilots, it is. It is my sincere hope that you find your place in aviation; that you serve it well; and that you become as safe and competent a pilot as is humanly possible. (See FIG. 11-11.)

A good friend of mine perhaps said it best when he said to me, "Wouldn't it be horrible if we had to punch a clock like so many other poor mortals do? God, what a life we live. I can't imagine not flying for a living." Nor could I, my friend. Nor could I.

So, if flight is your dream, go get it. Don't let anyone talk you out of it for any reason. If you desire it, it can happen. Good luck and God bless you.

Fig. 11-10. *Crop dusting is a career that offers excitement and good pay.*

Fig. 11-11. *Boeing's 747/400. Maybe, someday.*

Glossary

administrator The Federal Aviation Administrator or any person to whom he has delegated his authority in the matter concerned.

aerodynamic coefficients Nondimensional coefficients for aerodynamic forces and moments.

air carrier A person who undertakes directly by lease or other arrangement to engage in air transportation.

air commerce Interstate, overseas, or foreign air commerce, or the transportation of mail by aircraft or any operations or navigation of aircraft within the limits of any federal airway or any operation or navigation of aircraft that directly affects, or that may endanger safety in interstate, overseas, or foreign air commerce.

aircraft A device that is used or intended to be used for flight in the air.

aircraft engine An engine that is used or intended to be used for propelling aircraft. It includes turbosuperchargers and accessories necessary for its functioning, but does not include propellers.

airframe The fuselage, booms, nacelles, cowlings, fairings, airfoil surfaces (including rotors but excluding propellers and rotating airfoils of engines), and landing gear of an aircraft and their accessories and controls.

airplane A heavier-than-air, engine-driven, fixed-wing aircraft that is supported in flight by the dynamic reaction of the air against its wings.

airport An area of land or water that is used or intended to be used for the landing and takeoff of aircraft. Includes its buildings and facilities, if any.

GLOSSARY

airport traffic area (Unless otherwise specifically designated in FAR Part 93) that airspace within a horizontal radius of five statute miles from the geographical center of any airport at which a control tower is operating, extending from the surface up to, but not including, an altitude of 3,000 feet above the elevation of the airport.

airship An engine-driven, lighter-than-air aircraft that can be steered.

air traffic Aircraft operating in the air or on an airport surface, exclusive of loading ramps and parking areas.

air traffic clearance An authorization by air traffic control for the purpose of preventing collision between known aircraft and for an aircraft to proceed under specified traffic conditions within controlled airspace.

air traffic control A service operated by appropriate authority to promote the safe, orderly, and expeditious flow of air traffic.

air transportation Interstate, overseas, or foreign air transportation or the transportation of mail by aircraft.

alternate airport An airport at which an aircraft may land if a landing at the intended airport becomes inadvisable.

altitude engine A reciprocating aircraft engine having a rated takeoff power that is producible from sea level to an established higher altitude.

appliance Any instrument, mechanism, equipment, part, apparatus, appurtenance, or accessory (including communications equipment) that is used or intended to be used in operating or controlling an aircraft in flight and is installed in or attached to the aircraft but is not part of an airframe, engine, or propeller.

approved (Unless used with reference to another thing) approved by the administrator.

area navigation (RNAV) A method of navigation that permits aircraft operations on any desired course within the coverage of station-referenced navigation signals or within the limits of self-contained system capability.

area navigation high route An area navigation route within the airspace extending upward from and including 18,000 feet MSL to flight level 450.

area navigation low route An area navigation route within the airspace extending upward from 1,200 feet above the surface of the earth to, but not including, 18,000 feet MSL.

balloon A lighter-than-air aircraft that is not engine driven.

brake horsepower The power delivered at the propeller shaft (main drive or main output) of an aircraft engine.

calibrated airspeed Indicated airspeed of an aircraft, corrected for position and instrument error. Calibrated airspeed is equal to true airspeed in standard atmosphere at sea level.

category (1) As used with respect to the certification, rating, privileges, and limitations of airmen, means a broad classification of aircraft. Examples include: airplane; rotorcraft; glider; and lighter-than-air; and (2) as used with respect to the certification of aircraft, means a grouping of aircraft based upon intended use or operating limitation. Examples include: transport; normal; utility; acrobatic; limited; restricted; and provisional.

category II operation (With respect to the operation of aircraft), a straight-in ILS approach to the runway of an airport under a Category II ILS instrument approach procedure issued by the Administrator or other appropriate authority.

ceiling The height above the earth's surface of the lowest layer of clouds or obscuring phenomena that is reported as *broken, overcast,* or *obscuration,* and not classified as *thin* or *partial.*

civil aircraft Aircraft other than public aircraft.

class As used with respect to the certification, rating, privileges, and limitation of airmen, means a classification of aircraft within a category having similar operating characteristics. Examples include: single-engine; multiengine; land; water; gyroplane; helicopter; airship; and free balloon; and (2) as used with respect to the certification of aircraft, means a broad grouping of aircraft having similar characteristics of propulsion, flight, or landing. Examples include: airplane; rotorcraft; glider; balloon; landplane; and seaplane.

clearway (1) For turbine-engine-powered airplanes certificated after August 29, 1959, an area beyond the runway, not less than 500 feet wide, centrally located about the extended centerline of the runway, and under the control of the airport authorities. The clearway is expressed in terms of a clearway plane, extending from the end of the runway with an upward slope not exceeding 1.25 percent, above which no object nor any terrain protrudes. However, threshold lights might protrude above the plane if their height above the end of the runway is 26 inches or less and if they are located to each side of the runway; and (2) for turbine-engine-powered airplanes certificated after September 30, 1958, but before August 30, 1959, an area beyond the takeoff runway extending no less than 300 feet on either side of the extended centerline of the runway, at an elevation no higher than the elevation of the end of the runway, clear of all fixed obstacles, and under the control of the airport authorities.

commercial operator A person who, for compensation or hire, engages in the carriage of aircraft in air commerce of persons or property, other than as an air carrier or foreign air carrier or under the authority of Part 375 of this Title. Where it is doubtful that an operation is for *compensation or hire,* the test applied is whether the carriage by air is merely incidental to the person's other business or is, in itself, a major enterprise for profit.

controlled airspace Airspace designated as a continental control area, control area, control zone, terminal control area, or transition area, within which some or all aircraft may be subject to air traffic control.

crewmember A person assigned to perform duty in an aircraft during flight time.

critical altitude The maximum altitude at which, in standard atmosphere, it is possible to maintain, at a specified rotational speed, a specified power or a specified manifold pressure. Unless otherwise stated, the critical altitude is the maximum altitude at which it is possible to maintain, at the maximum continuous rotational speed, one of the following: (1) the maximum continuous power, in the case of engines for which this power rating is the same at sea level and at the rated altitude; and (2) the maximum continuous rated manifold pressure, in the case of engines, the maximum continuous power of which is governed by a constant manifold pressure.

GLOSSARY

critical engine The engine whose failure would most adversely affect the performance of handling qualities of an aircraft.

decision height (With respect to the operation of aircraft) the height at which a decision must be made, during an ILS or PAR instrument approach, to either continue the approach or to execute a missed approach.

equivalent airspeed The calibrated airspeed of an aircraft corrected for adiabatic compressible flow for the particular altitude. Equivalent airspeed is equal to calibrated airspeed in standard atmosphere at sea level.

extended over-water operation (1) With respect to aircraft other than helicopters, an operation over water at a horizontal distance of more than 50 nautical miles from the nearest shoreline; and (2) with respect to helicopters, an operation over water at a horizontal distance of more than 50 nautical miles from the nearest shoreline and more than 50 nautical miles from an off-shore heliport structure.

external load A load that is carried or extends outside of the aircraft fuselage.

external-load attaching means The structural components used to attach an external load to an aircraft, including external-load container, the backup structure at the attachment points, and any quick-release device used to jettison the external load.

fireproof (1) With respect to materials and parts used to confine fire in a designated fire zone, the capacity to withstand—at least as well as steel in dimensions appropriate for the purpose for which they are used—the heat produced when there is a severe fire of extended duration in that zone; and (2) with respect to other materials and parts, the capacity to withstand the heat associated with fire at least as well as steel in dimensions appropriate for the purpose for which they are used.

fire resistant (1) With respect to sheet or structural members, the capacity to withstand the heat associated with fire at least as well as aluminum alloy in dimensions appropriate for the purpose for which they are used; and (2) with respect to fluid-carrying lines, fluid system parts, wiring, air ducts, fittings, and powerplant controls, the capacity to perform the intended functions under the heat and other conditions likely to occur when there is a fire at the place concerned.

flame resistant Not susceptible to combustion to the point of propagating a flame, beyond safe limits, after the ignition source is removed.

flammable With respect to fluid or gas, susceptible to igniting readily or exploding.

flap extended speed The highest speed permissible with wing flaps in a prescribed extended position.

flash resistant Not susceptible to burning violently when ignited.

flight crewmember A pilot, flight engineer, or flight navigator assigned to duty in an aircraft during flight time.

flight level A level of constant atmospheric pressure related to a reference datum of 29.92 inches of mercury stated in three digits that represent hundreds of feet. For example, flight level 250 represents a barometric altimeter indication of 25,000 feet; flight level 255, an indication of 25,000 feet.

flight plan Specified information relating to the intended flight of an aircraft that is filed orally or in writing with air traffic control.

flight time The time from the moment the aircraft first moves under its own power for the purpose of flight until the moment it comes to rest at the next point of landing-*block-to-block time.*

flight visibility The average forward horizontal distance from the cockpit of an aircraft in flight, at which prominent unlighted objects can be seen and identified by day and prominent lighted objects can be seen and identified by night.

foreign air carrier Any person other than a citizen of the United States who undertakes directly, by lease or other arrangement, to engage in air transportation.

foreign air commerce The carriage by aircraft of persons or property for compensation or hire, or the carriage of mail by aircraft, or the operation or navigation of aircraft in the conduct or furtherance of a business or vocation, in commerce between a place in the United States and any place outside thereof, whether such commerce moves wholly by aircraft or partly by aircraft and partly by other forms of transportation.

foreign air transportation The carriage by aircraft of persons or property as a common carrier for compensation or hire, or the carriage of mail by aircraft, in commerce between a place in the United States and any place outside the United States, whether that commerce moves wholly by aircraft or partly by aircraft and partly by other forms of transportation.

glider A heavier-than-air aircraft that is supported in flight by the dynamic reaction of the air against its lifting surfaces and whose free flight does not depend principally on an engine.

GPS Global Positioning System is a form of navigation using triangulation of signals from satellite transmitters to receivers in an aircraft.

ground visibility Prevailing horizontal visibility near the earth's surface as reported by the United States National Weather Service or an accredited observer.

gyrodine A rotorcraft whose rotors are normally engine-driven for taking off, hovering, and landing as well as for forward flight through part of its speed range. Means of propulsion, consisting usually of conventional propellers, are independent of the rotor system.

gyroplane A rotorcraft whose rotors are not engine-driven except for initial starting, but are made to rotate by action of the air when the rotorcraft is moving; and whose means of propulsion, consisting usually of conventional propellers, is independent of the rotor system.

helicopter A rotorcraft that for its horizontal motion depends principally on its engine-driven rotors.

heliport An area of land, water, or structure used or intended to be used for the landing and taking off of helicopters.

idle thrust The jet thrust obtained with the engine power control lever set at the stop for the least-thrust position at which it can be placed.

IFR conditions Weather conditions below the minimum for flight under visual flight rules.

IFR over-the-top With respect to the operation of aircraft, the operation of an aircraft over-the-top on an IFR flight plan when cleared by air traffic control to maintain *VFR conditions* or *VFR conditions on top.*

GLOSSARY

indicated airspeed The speed of an aircraft as shown on its pitot static airspeed indicator calibrated to reflect standard atmosphere adiabatic compressible flow at sea level uncorrected for airspeed system errors.

instrument A device using an internal mechanism to show visually or aurally the attitude, altitude, or operation of an aircraft or aircraft part. It includes electronic devices for automatically controlling an aircraft in flight.

interstate air commerce The carriage by aircraft of persons or property for compensation or hire, or the carriage of mail by aircraft, or the operation or navigation of aircraft in the conduct or furtherance of a business or vocation, in commerce between a place in any state of the United States or the District of Columbia, and a place in any other state of the United States or the District of Columbia; or between places in the same state of the United States through the airspace over any place outside thereof; or between places in the same territory or possession of the United States or the District of Columbia.

interstate air transportation The carriage by aircraft of persons or property as a common carrier for compensation or hire, or the carriage of mail by aircraft, in commerce (1) between a place in the state or the District of Columbia and another place in the state or the District of Columbia; (2) between places in the same state through the airspace of any place outside that state; or (3) between places in the same possession of the United States; whether that commerce moves wholly by aircraft or partly by aircraft and partly by other forms of transportation.

intrastate air transportation The carriage of persons or property as a common carrier for compensation or hire, by turbojet-powered aircraft capable of carrying 30 or more persons, wholly within the same State of the United States.

landing gear extended speed The maximum speed at which an aircraft can be safely flown with the landing gear extended.

landing gear operating speed The maximum speed at which the landing gear can be safely extended or retracted.

large aircraft Aircraft of more than 12,500 pounds, maximum certificated takeoff weight.

lighter-than-air aircraft Aircraft that can rise and remain suspended by using contained gas weighing less than the air that is displaced by the gas.

load factor The ratio of a specified load to the total weight of the aircraft. The specified load is expressed in terms of any of the following: aerodynamic forces, inertia forces, or ground or water reactions.

LORAN The acronym for LOng RAnge Navigation. The form of navigation based on the timing of electronic impulses from transmitters on earth to a receiver in an aircraft.

mach number The ratio of true airspeed to the speed of sound.

maintenance Inspection, overhaul, repair, preservation, and the replacement of parts, but excludes preventive maintenance.

major alteration An alteration not listed in the aircraft, aircraft engine, or propeller specifications (1) that might appreciably affect weight, balance, structural strength, performance, powerplant operation, flight characteristics, or other qualities affecting airworthiness; or (2) that is not done according to accepted practices or cannot be done by elementary operations.

major repair A repair (1) that, if improperly done, might appreciably affect weight, balance, structural strength, performance, powerplant operation, flight characteristics, or other qualities affecting airworthiness; or (2) that is not done according to accepted practices or cannot be done by elementary operations.

manifold pressure Absolute pressure as measured at the appropriate point in the induction system and usually expressed by inches of mercury.

medical certificate Acceptable evidence of physical fitness on a form prescribed by the Administrator.

minimum descent altitude The lowest altitude, expressed in feet above mean sea level, to which descent is authorized on final approach or during circle-to-land maneuvering in execution of a standard instrument approach procedure, where no electronic glide slope is provided.

minor alteration An alteration other than a major alteration.

minor repair Repair other than a major repair.

navigable airspace Airspace at and above the minimum flight altitudes prescribed by regulations, including airspace needed for safe takeoff and landing.

night The time between the end of evening civil twilight and the beginning of morning civil twilight, as published in the American Air Almanac, converted to local time.

nonprecision approach procedure A standard instrument approach procedure in which no electronic glide slope is provided.

operate With respect to aircraft, means use, cause to use, or authorize to use aircraft for the purpose (except as provided in FAR 91) of air navigation including the piloting of aircraft, with or without the fight of legal control (as owner, lessee, or otherwise).

operational control With respect to a flight, the exercise of authority over initiating, conducting, or terminating a flight.

overseas air commerce The carriage by aircraft of persons or property for compensation or hire, or the carriage of mail by aircraft, or the operation or navigation of aircraft in the conduct or furtherance of a business or vocation, in commerce between a place in any state of the United States or the District of Columbia, and any place in a territory or possession of the United States; or between a place in a territory or possession of the United States and a place in any other territory or possession of the United States.

overseas air transportation The carriage by aircraft of persons or property as a common carrier for compensation or hire, or the carriage of mail by aircraft, in commerce (1) between a place in a state or the District of Columbia and a place in a possession of the United States; or (2) between a place in a possession of the United States and a place in another possession of the United States; whether the commerce moves wholly by aircraft or partly by aircraft and partly by other forms of transportation.

over-the-top Above the layer of clouds or other obscuring phenomena forming the ceiling.

GLOSSARY

person An individual, firm, partnership, corporation, company, association, joint-stock association, or governmental entity. It includes a trustee, receiver, assignee, or similar representative of any of them.

pilotage Navigation by visual reference to landmarks.

pilot in command The pilot responsible for the operation and safety of an aircraft during flight time.

pitch setting The propeller blade setting as determined by the blade angle measured in a manner, and at a radius, specified by the instruction manual for the propeller.

positive control Control of all air traffic, within designated airspace, by air traffic control.

precision approach procedure Standard instrument approach procedure in which an electronic glide slope is provided, such as ILS and PAR.

preventive maintenance Simple or minor preservation operations and the replacement of small standard parts not involving complex assembly operation.

prohibited area Designated airspace within which the flight or aircraft is prohibited.

propeller A device for propelling an aircraft that has blades on an engine-driven shaft and that, when rotated, produces by its action on the air a thrust approximately perpendicular to its plane of rotation. It includes control components normally supplied by its manufacturer, but does not include main and auxiliary rotors or rotating airfoils of engines.

public aircraft Aircraft used only in the service of a government or a political subdivision. It does not include any government-owned aircraft engaged in carrying persons or property for commercial purposes.

rated takeoff power (With respect to reciprocating, turbopropeller, and turboshaft engine type certification) the approved brake horsepower that is developed statically under standard sea level conditions, within the engine operating limitations established under Part 33, and limited in use to periods of not over five minutes for takeoff operation.

rating A statement that, as a part of a certificate, sets forth special conditions, privileges, or limitation.

reporting point A geographical location in relation to which the position of an aircraft is reported.

restricted area Airspace designated within which the flight of aircraft, while not wholly prohibited, is subject to restriction.

RNAV waypoint (W/P) A predetermined geographical position used for route or instrument approach defined for progress-reporting purposes that is defined relative to a VORTAC station position.

rocket An aircraft propelled by ejected expanding gases generated in the engine from self-contained propellants and not dependent on the intake of outside substances. It includes any part that becomes separated during the operation.

rotorcraft A heavier-than-air aircraft that depends principally for its support in flight on the lift generated by one or more rotors.

route segment A part of a route. Each end of that part is identified by (1) a continental or insular geographical location; or (2) a point at which a definite radio fix can be established.

sea-level engine A reciprocating aircraft engine having a rated takeoff power that is producible only at sea level.

second in command A pilot who is designated to be second in command of an aircraft during flight time.

show (Unless the context otherwise requires) to show to the satisfaction of the Administrator.

small aircraft Aircraft of 12,500 pounds or less maximum certification takeoff weight.

standard atmosphere The atmosphere defined in *U.S. Standard Atmosphere, 1962* (Geopotential altitude tables).

stopway An area beyond the takeoff runway, no less wide than the runway and centered upon the extended centerline of the runway, able to support the airplane during an aborted takeoff without causing structural damage to the airplane and designated by the airport authorities for use in decelerating the airplane during an aborted takeoff.

takeoff power (1) With respect to reciprocating engines, the brake horsepower that is developed under standard sea-level conditions, and under the maximum conditions of crankshaft rotational speed and engine manifold pressure approved for the normal takeoff and limited in continuous use to the period of time shown in the approved engine specification; and (2) with respect to turbine engines, the brake horsepower that is developed under static conditions at a specified altitude and atmospheric temperature, and under the maximum conditions of rotorshaft rotational speed and gas temperature approved for the normal takeoff, and limited in continuous use to the period of time shown in the approved engine specification.

time in service With respect to maintenance time records, the time from the moment an aircraft leaves the surface of the earth until it touches it at the next point of landing.

traffic pattern The traffic flow that is prescribed for aircraft landing at, taxiing on, or taking off from an airport.

true airspeed The airspeed of an aircraft relative to undisturbed air.

type (1) As used with respect to the certification, rating, privileges, and limitations of airmen, a specific make and basic model of aircraft, including modifications thereto that do not change its handling or flight characteristics. Examples include: DC-10, 1049, and F-14; (2) as used with respect to the certification of aircraft, those aircraft that are similar in design. Examples include: DC-7 and DC-7C; 1049G and 1049H; and F-27 and F-27F; and (3) as used with respect to the certification of aircraft engines, those engines that are similar in design. For example, JT8D and JT8D-7 are engines of the same type and JT9D-3A and JT9D-7 are engines of the same type.

United States (In a geographical sense) (1) the states, the District of Columbia, Puerto Rico, and the possessions, including the territorial waters; and (2) the airspace of those areas.

United States Air Carrier A citizen of the United States who undertakes directly by lease or other arrangement to engage in air transportation.

VFR over-the-top With respect to the operation of aircraft, the operation of an aircraft over-the-top under VFR when it is not being operated on an IFR flight plan.

V speeds In general, the reference-indicated airspeeds used to limit an aircraft's performance.

GLOSSARY

V_a Design maneuvering speed. The maximum speed at which the pilot may use full, abrupt control travel without causing structural damage to the aircraft.

V_{fe} Maximum flap extension speed. Shown at the top (highest speed) of the white arc on the airspeed indicator.

V_{le} Maximum landing gear extension speed.

V_{lo} Maximum speed to operate the landing gear up or down.

V_{lof} Liftoff speed.

V_{mc} The minimum airspeed at which it's possible to maintain directional control of the aircraft within 20 degrees of heading and, thereafter, maintain straight flight with not more than 5 degrees of bank if one engine fails suddenly.

V_{ne} The maximum speed at which an aircraft may be operated under any circumstances.

V_{no} Maximum velocity for normal operations found at the top of the green arc on the airspeed indicator.

V_r Rotation airspeed.

V_{sse} Minimum speed at which to purposely fail an engine during flight.

V_{so} Stall speed in the landing configuration with landing gear and flaps down and power at idle. Is shown on the airspeed indicator as the lowest number on the white arc.

V_{s1} Stall speed in the clean configuration with landing gear and flaps up and power at idle.

V_x Best angle of climb airspeed. This airspeed gives the aircraft the greatest gain of altitude in a given distance.

V_{xse} Best angle of climb-single-engine. This airspeed gives the aircraft the greatest gain of altitude in a given distance when operating on a single engine.

V_y Best rate of climb airspeed. This airspeed gives the aircraft the greatest gain of altitude in a given period of time.

V_{yse} Best rate of climb airspeed. This airspeed gives the aircraft the greatest gain of altitude in a given period of time when operating on a single engine.

Index

INDEX

INDEX

About the Author

David A. Frazier (Jackson, MI) is currently Director of Aviation at Jackson College; prior to this, he taught aviation at Vincennes University (Indiana) for 20 years. He is a multi-rated pilot with over 10,000 hours of instructing experience, in both aircraft and simulators. Frazier has been an FAA Designated Pilot Examiner since 1982. He is the author of *AG Pilot flight Training Guide*, and *How to Master Precision Flight*.